Praise for *The Arbornaut*

"*The Arbornaut* is about a shy girl who loved to play outdoors and became a scientist who educated the world about the abundant life in the treetops. I loved it."

—TEMPLE GRANDIN, author of *Animals Make Us Human*

"*The Arbornaut* is an eye-opening and enchanting book by one of our major scientist-explorers. Meg Lowman is the perfect guide to a little-known, life-besotted world, and she has fascinating stories to tell about her life in the top story."

—DIANE ACKERMAN, author of *The Zookeeper's Wife*

"Meg Lowman—'CanopyMeg'—is my true hero, a courageous explorer who makes amazing discoveries high in the forest canopy. *The Arbornaut* captures the magic of that little-known world with its pioneering stories and clear, informative text. Readers everywhere will be fascinated and inspired to learn more about nature, and especially about how we need to conserve the world's forests."

—JANE GOODALL, PhD, DBE, founder of the Jane Goodall Institute and United Nations Messenger of Peace

"Perfect for anyone seeking to deepen their knowledge of trees, forest, and the remarkable interconnectedness of nature."

—ERIC LIEBETRAU, *Kirkus Reviews* (starred review)

"Full of life, energy, intelligence and determination . . . A book to reach for if you, like Lowman, love the natural world and want to live in it fully." —KELLY BLEWETT, *BookPage* (starred review)

"*The Arbornaut* [is] about an extraordinary life, one spent among trees."

—SOPHIE CUNNINGHAM, *The Guardian*

"Lowman's contributions to ecology are numerous, as a professor, science communicator, international collaborator, and leader in conservation organizations. In this science-oriented memoir, she details a lifetime of experiences . . . A highly engaging read."

—CATHERINE LANTZ, *Library Journal* (starred review)

"A passionate look at the 'unexplored wonderland' of trees . . . Nature lovers will find much to consider."　　　　—*Publishers Weekly*

"Written . . . not just to instruct, but to reorient and inspire . . . If a tree was once understood as a mostly static living object, [through *The Arbornaut*] we see it rippling with change."

—REBECCA GIGGS, *The Atlantic*

"A fascinating journey."　　　　　　　—SARAH BOON, *Undark*

"Lowman details a career that embodies the word 'trail-blazing' . . . From start to finish, the word that best describes *The Arbornaut* is 'spirited.'"　　　—EMILY DONALDSON, *The Globe and Mail* (Toronto)

"Amazing . . . This is an inspiring book for any young woman considering a career in the biological sciences and [one that] encourages all of us to become 'citizen scientists' that are stewards of big trees."

—WICN Public Radio (Worcester, MA)

"An ideal book for city-bound nature lovers suffering withdrawal during lockdown . . . [Lowman] forged a remarkable career researching canopies all over the world and educating the public on the wonders of this 'eighth continent' in the tree tops."

—*The Sydney Morning Herald*

"[Lowman] tells her story with remarkable detail, highlighting her research methods and discoveries as well as the adversity she faced as a woman in the sciences and a single mother."

—CARLYN KRANKING, *Audubon* magazine

"Everyone will want to read this book. Meg Lowman is starting a whole new movement exploring the treetops!"

—E. O. WILSON, author of *Half-Earth: Our Planet's Fight for Life*

"An exciting, firsthand introduction to some of the things we must come to understand if we are to secure a sustainable future for humanity in these challenging times. A thoroughly enjoyable yet deeply meaningful contribution in every respect!"

—PETER H. RAVEN, PhD, president emeritus of the Missouri Botanical Garden

"An account of intrepid exploration at the upper reaches of terrestrial life, where branches and foliage touch the sky and all creatures awake to the first morning rays of the sun."

—WADE DAVIS, author of *Magdalena, River of Dreams: A Story of Colombia*

"In *The Arbornaut*, Meg Lowman inspires readers with her amazing life story and reminds us that we can't live without the natural world, especially trees."

—TOMMY HILFIGER

MEG LOWMAN

The Arbornaut

Meg Lowman, PhD, aka "CanopyMeg," is an American biologist, educator, ecologist, writer, editor, and public speaker. She is the executive director of the TREE Foundation and a professor at the National University of Singapore, Arizona State University, and Universiti Sains Malaysia. Nicknamed the "real-life Lorax" by *National Geographic* and "Einstein of the treetops" by *The Wall Street Journal*, Lowman pioneered the science of canopy ecology. Her motto is "no child left indoors." She travels extensively for research, outreach, and speaking engagements for audiences large and small.

The Arbornaut

A

LIFE DISCOVERING

THE

EIGHTH CONTINENT

IN THE

TREES ABOVE US

Meg Lowman

PICADOR

FARRAR, STRAUS AND GIROUX

NEW YORK

Picador
120 Broadway, New York 10271

Printed in the United States of America
Originally published in 2021 by Farrar, Straus and Giroux
First paperback edition, 2022

The Library of Congress has cataloged the Farrar, Straus and Giroux hardcover
edition as follows:
Names: Lowman, Margaret, author.
Title: The arbornaut: a life discovering the eighth continent in the trees above us /
   Meg Lowman.
Description: First edition. | New York: Farrar, Straus and Giroux, 2021. |
   Includes index.
Identifiers: LCCN 2021010520 | ISBN 9780374162696 (hardcover)
Subjects: LCSH: Lowman, Margaret. | Plant ecologists—United States—
   Biography. | Botanists—United States—Biography. | Rain forest ecology—
   Research. | Forest canopy ecology—Research.
Classification: LCC QK31.L69 A3 2021 | DDC 581.7092 [B]—dc23
LC record available at https://lccn.loc.gov/2021010520

Paperback ISBN: 978-1-250-84918-2

Designed by Abby Kagan

This book is dedicated to those lifelong planetary heroes—trees. I hope my passion for these leafy giants will inspire readers to share a sense of wonder for our eighth continent, and maybe we can help save it, together.
Big thanks to Eddie and James for joyfully climbing in many forests with their mom.

CONTENTS

※》《※

FOREWORD
※※

I WILL NEVER LOOK AT A TREE in the same way again, nor will the rest of the world, thanks to the author of this book. It now seems obvious that most of what makes a tree—as well as what constitutes a forest—is above eye level, but until Meg Lowman's irrepressible curiosity inspired her to look at trees from the top down, most humans tended to view them from the bottom up. What they missed is most of what makes treetops not only individual miracles, but collectively the source of shelter and sustenance for most forest dwellers, with dividends for the rest of life on Earth. I was thrilled when I heard about a fellow botanist who had devised ways to not only use her natural primate abilities to climb trees, but also to take tree-climbing to new heights with ingenious lifting techniques and, pushing further, to develop sky-walking pathways among the trees' leafy crowns. In this engaging volume, she shares her view with stories that you know are true because "You just can't make this stuff up!"

For a scientist and explorer, it is satisfying to make new discoveries, to go where no woman (or man) has gone before, to see what others have not, and to find meaningful pieces of the great living puzzle of life that is unique to Earth. But Meg Lowman does more, excelling in communicating her findings not only to scientists in the arcane

language of numbers and graphs, but also to non-scientists with con-
tagious enthusiasm and meaningful rationale, in language and humor
befitting the audience, conveying why trees matter and how their existence and ours is inextricably connected. She also instills a sense of
urgency about embracing the planet's remaining natural forests with
enhanced protection, speaking in classrooms and boardrooms, in villages far from tall buildings, in the offices of government officials,
electronically and in print, with the world at large.

Throughout history, people have taken from nature whatever was
needed or wanted from the world's lands and waters. When our numbers were small and the natural world was largely intact, our impact
was slight, but after one hundred thousand years of a more or less
peaceful relationship with nature, the past five hundred, and especially
the past fifty, have marked a turning point that does not bode well
for the future of life on Earth. Human capacity to consume and alter
the nature of nature has reached perilous tipping points for climate,
biodiversity, and land and water use, compounded by pollution, all
driving changes in planetary processes and the underpinnings of what
makes Earth hospitable for life as we know it. The good news is the
other tipping point—knowledge. Children in the twenty-first century
(adults, too) are armed with the superpower of knowing what Earth
looks like from space, of seeing and hearing about events across the
globe in real time while understanding the new perspectives of geological time, of seeing Earth's place in the universe, of vicariously
traveling into the inner workings of cells, to the depths of the deepest seas, and to the tops of the highest trees. Half a century ago, it was
widely believed that Earth was too big to fail. Now we know. If Earth
is to remain habitable for the likes of us, we must take care of what remains of the natural systems that took 4.5 billion years to make and a
bit more than 4.5 decades to break and do our best to restore damaged
areas to better health. There is still time to hold on to the last safe
havens where trees are intact, hosting miraculous creatures that are as
vital to our existence as we are to theirs.

Bravo, Meg Lowman, aka "Your Highness," for sharing your journey in this book, and for launching Mission Green, thereby inspiring others to understand and know why we must take care of the natural world as if our lives depend on it. Because they do.

    —SYLVIA A. EARLE, aka "Her Deepness," founder, Mission Blue Oceanographer, botanist, National Geographic Explorer in Residence

TEN TIPS OF FIELD BIOLOGY FOR EVERY
ASPIRING ARBORNAUT

1. Always carry a headlamp, not just in the forest but anywhere . . . even on a plane or traveling by car.
2. Keep a few tissues in your pocket for those emergency ablutions behind a tree!
3. Wear a vest with lots of pockets.
4. Never drink more than half of your water supply every time you hydrate so you always have some left, and it always helps to tell someone your itinerary in case a rescue is required.
5. Keep your camera handy for amazing discoveries, even if it is just on your phone.
6. Carry a poncho. It can serve as a ground cloth as well as rain gear.
7. Oreo cookies are a wonderful energy snack!
8. If you are a parent, carry a few photos of your kids—they work well to break the ice with other cultures, especially where language barriers exist.
9. Use all five senses, relentlessly.
10. Keep a journal so that you can recall amazing stories, biodiversity, and observations.

The
Arbornaut

PROLOGUE

How to See the Whole Tree (and What That Means for the Forest)

IMAGINE GOING TO THE DOCTOR for a complete checkup and, in the course of an entire visit, the only body part examined was your big toe. The visit ends with a pronouncement that you are perfectly healthy, but there was no test of your vital signs, heartbeat, vision, or any other part of you—just the big toe. You may have gone in with a broken arm or a headache from high blood pressure, but the assessment of your lowest bipedal extremity alone couldn't clue the doctor in to the real trouble. How would you feel? At the very least, you'd probably switch doctors.

For centuries, the health of trees, even those ancient giants stretching hundreds of feet high into the clouds, was assessed in just the same way. Examining woody trunks at eye level, scientists essentially inspected the "big toes" of their patients and then made sweeping deductions about forest health without ever gazing at the bulk of the tree, known as the canopy, growing overhead. The only time foresters had the chance to evaluate a whole tree was when it was cut down—which is kind of like assessing a person's entire medical history from a few ashes after cremation. In tropical forests especially, the lower levels are as different from the upper reaches as night and day. The ground receives as little as 1 percent of the light shining on the crowns. So the understory is dark, windless, and often humid whereas the canopy is blasted with sun, whipped by high winds, and often crispy in its dryness between rainstorms. The gloomy forest floor is inhabited by a few shade-loving creatures, while the canopy hosts a riotous variety of life—millions of species of every imaginable color, shape, and size that pollinate flowers, eat leaves, and also eat each other.

Before the 1980s, foresters unimaginatively overlooked 95 percent of their subject; almost no one paid attention to the treetops. Then, in 1978, a young botanist with a lifelong passion for green giants and infatuated by leaves arrived in Australia on a fellowship to study tropical forests. Coming from the temperate zones, this neophyte knew almost nothing about the tropics. During her first visit to a rain forest in Australia, she stared up into the most dizzyingly tall trees she'd ever met and thought, "Holy cow, I can't see the top!" That gobsmacked botanist was me.

I carried with me an enormous tree-love and planned to devote my future to demystifying their secrets. After a few misadventures, I realized that to understand the whole forest, I needed to get up into its highest levels. Initially I hoped that the simple use of binoculars would be enough to bring the treetops down to me. But after a lot of thought and some trial and error, I figured out a way to hoist myself into that magical, unexplored wonderland, full of the six-legged mayhem of the insect world and more shades of green than I imagined possible. I nicknamed this amazing new world the "eighth continent." Cavers go down a rope, but I went up. Recreational mountaineers pound hardware into rock cliffs, but I gently rigged tall trees to avoid breaking any leaves or scaring off any creatures. And to affix my ropes in the upper branches, I welded my own slingshot from a metal rod. My approach turned out to be a simple, inexpensive technique, and it launched my exploration of that "eighth continent," a complex hot spot of biodiversity located not hundreds or thousands of miles away like the ocean floor or outer space, but almost within reach just above our heads. I called myself an "arbornaut."

During that first ascent into the canopy, I ecstatically came eyeball to eyeball with creatures I had never imagined and that were, at that very moment, still unknown to the rest of the world. I marveled at a handsome black-snouted weevil sucking leaf juices, elegant colorful pollinators flitting between vine flowers, giant bird-nest ferns that gave sanctuary to ants, and thousands upon thousands of my favorite thing: leaves. As I moved from bottom to top, I was dumbstruck by the changes I observed. Foliage in the shaded understory was blackish-green, larger, thinner, and, as it turned out, more long-lived (thanks to the windless, protected, and dark environment near the forest floor). Leaves in bright sunshine at the top were small, leathery, yellowish green, and very tough. Everywhere I looked, the crowns shared secrets not visible from ground level—shiny beetles ate young (but not old) leaf tissue, caterpillars operated in gangs feeding on entire branches from youngest to older foliage, birds plucked these unsuspecting larvae as if feasting at a salad bar, and sudden downpours of rain sent all the critters scrambling for shelter under the nearest foliage or bark crevice. In the years ahead, treetop exploration would lead to the discovery that upward of *half* of all terrestrial creatures live about one hundred feet or more above our heads, not at ground level as scientists had previously assumed. As I soon discovered, in the upper crowns, the majority of species were new to science. Across more than sixty thousand species of trees, nearly every one hosts unique communities.

When they encounter unexplored frontiers, scientists design new techniques and technology for safe exploration. The invention of the self-contained underwater breathing apparatus, or SCUBA, in the 1950s opened the extraordinary world of coral reef biodiversity to scientific research. During the 1960s, astronauts landed on the moon thanks to NASA's development of rocket combustion for space travel. Solid rocket fuel was to astronauts what my humble homemade slingshot was to arbornauts—not a new invention, but an innovative new way to use one. Just as space travel launched a generation of astronauts, canopy access created a new career pathway for arbornauts. If you love to climb trees, take note—there is a profession for you! I was one of those first arboreal explorers, and arguably the only crazy climber to have conducted research on every continent (even Antarctica, where the tops of moss and lichen foliage are only two inches high, and require kneeling, not climbing, to access their Lilliputian canopies). Over the past forty years, I have marked thousands of leaves and tracked their life history, some lasting more than twenty years despite the constant threats of creatures (mostly insects) trying to eat, tear, tunnel through, or disfigure them. And this aerial shift in our approach to forest science has also led to advancements in knowledge of global cycles ranging from freshwater circulation to carbon storage to climate change.

It shouldn't be a surprise (but it still is for some) that planetary health links directly to forests. Their canopies produce oxygen, filter fresh water, transfer sunlight into sugars, clean our air by absorbing carbon dioxide, and provide a home to the extraordinary genetic library of all earthbound creatures, among many other crucial functions. And unlike an electric grid or water treatment plant, no expensive taxes or fees are required to maintain this complex forest machinery that keeps our Earth healthy. Still, for it to function well, we need to insulate it from human destruction. Within my lifetime of approximately six decades, Amazon rain forest degradation has zoomed past a tipping point; restoration is unlikely. Countries like Madagascar, Ethiopia, and the Philippines have almost no primary forests left to seed future stands. And remaining forest fragments around the world ranging from California to Indonesia to Brazil are at great risk from fires, drought, roads, and clearing. We must race ever faster to understand the mysteries of the treetops before they disappear, or better yet, we must find a way to conserve those remaining green Noah's arks. "Climate change" was not in my vocabulary when I studied backyard trees some fifty years ago, but now this term drives an even greater urgency to understand and conserve natural systems, especially forests.

One way to save more trees is to introduce more people to their wonders. After perfecting safe rope techniques, I designed aerial trails in the canopy called treetop walkways or skywalks, allowing groups of people to study the crowns, not just one person dangling from a rope. These walkways not only offered an important research and education tool but also served a humanitarian effort; they provided income to indigenous people from ecotourism instead of logging, which in turn inspires sustainable conservation. After ropes and walkways, I went on to design, tinker with, and utilize an extensive toolkit for canopy exploration, including cherry pickers, hot-air balloons, construction cranes, and drones. Each tool allowed unique access to different aspects of the forest and provided the means to answer diverse research questions. Exploration of whole forests, not merely the forest floor, has inspired communities around the globe—from governments in Malaysia to priests in Ethiopia—to partner with arbornauts to save their precious green heritage, which is so critical to human survival. In my experience, positive conservation outcomes arose more from mutual trust between scientists and local stakeholders than from the latest technical publications. And it never hurts to invite a few community leaders into the canopy! People seem to love climbing trees—even those who think they've outgrown it.

No one would have guessed that a shy kid from rural upstate New York, a veritable geek who spent her childhood collecting wildflowers along roadsides, could change our view of the planet with a few home-made gadgets. Using a simple toolkit that fits into one duffel bag, I now travel the world exploring the eighth continent, uncovering its secrets, and sharing treetop wonders with anyone who will listen. My story is a testimony: any average kid can make discoveries by exploring the world around us. This book is meant to share that thrill of aerial exploration—filled with thousands of feet of ropes, many failed volleys with a welded slingshot, lots of remote jungles, hundreds of thousands of leaves examined at their birthplace (not clipped or dead on the forest floor), multitudes of stinging ants, and gazillions of other creatures in their green penthouses. After four decades as an arbornaut, forests remain my best teachers. Once you have ascended into the canopy with me through these pages, I'm betting you, too, will feel an urgency to champion their conservation.

1

FROM WILDFLOWER TO WALLFLOWER
⇉⥿
A Girl Naturalist in Rural America

I HAD SET THE ALARM, but in my excitement, I awoke a half hour before it rang. At 4:00 a.m., there were still just slivers of light on the horizon, so I tiptoed out to the living room of our small cabin on Seneca Lake to avoid waking my two younger brothers.

Our town of Elmira, New York, was unbearably hot in the summer, so we escaped to a nearby cabin twenty-five miles away, where the forests worked their natural cooling magic. The cottage where we spent these summers was an abandoned gristmill. Many years before, probably almost a hundred, an elm tree had taken root on that mill site; my grandfather, a stonemason and carpenter, then lovingly constructed the cottage around its trunk, so it ran right up through the living room as a prominent feature. When it rained, water dripped through the roof, down the bark to a patch of soil nestled amid the stone floor. I often searched for tiny insects inhabiting all the woody fissures. The elm canopy stretched across the entire roof, shading our cottage during summer and standing sentinel with its bare branches in winter. I loved every inch of that tree right down to the diverse fungi that grew along the trunk when it fell victim to Dutch elm disease. Its special place inside our cabin was always such a comfort, and one of my saddest childhood moments was its death. On a dangerously tall ladder, my grandfather meticulously cut away all the dead branches, leaving the beloved trunk as a statue gracing the center of the cottage. My grandparents would occasionally allow me to harvest one of the flat, tough fungal brackets decorating the dead trunk adjacent to our dining room table. I did my creative best to etch designs and paint images of plants on these living canvases, sometimes called artist fungus. Summer at our rustic cabin was my refuge, where I could explore, observe, and collect, inhaling all the nature my small body could absorb.

I carefully shook Mom awake, then we crept outside before sunrise and quietly drove five miles along a dirt road to my favorite birdwatching pond. This nature trip was a huge deal for me. Mom did not even own binoculars, and all she knew about my feathered friends was that starlings sometimes tore up her spring lettuce sprouts. But she knew birdwatching brought immeasurable joy to me at the tender age of seven and so offered to drive to a special place for the dawn chorus—an exquisite concert sung by a winged choir at sunrise. Our old Rambler bounced along a dusty road, through cherry orchards where my brother and I earned money picking fruit, near an old haunted house that made my hair stand on end, past a one-room bar where farmers bragged about their corn. Willows encircled the pond's edge, their roots adapted to soggy soil. A discarded, leaky old rowboat was tied to a branch. I had dreamed all summer of rowing out from the bank to observe egrets or herons. Those majestic birds would garner five big stars on my modest bird list if they appeared. When we arrived, my mom worried we were trespassing on some farmer's property even though there was no house to be seen, so she very reluctantly climbed into that dilapidated dinghy, strewn with spiderwebs and dusty underpinnings. Away we paddled. This was as close as I ever came to feeling like a princess in a silver carriage, even though it was a rough-and-tumble contraption full of leaks! We rowed and bailed constantly, to keep afloat. Out in the middle of the pond, we stopped paddling and I focused my enormous Sears binoculars. They were ridiculously bulky, felt like they weighed almost as much as I did, and probably were not even capable of focusing, but they made me feel like a professional ornithologist. Much to my amazement, as if on cue, a great blue heron flew in and landed along the shoreline. Even my mom was overcome with awe.

Whereas kids today are beset with indoor technologies and fall prey to screen fatigue, I probably suffered from an overdose of plant oxygen and green blindness. From the day I could walk, I was a tireless collector of natural stuff. At the lake, I amassed piles of special shells and stones. Back in Elmira under my bed, I stashed wildflowers, twigs, bird nests, more stones, feathers, dead twigs (to study winter buds!), and even snakeskins. My parents indulged my love for nature in small but thoughtful ways, always stopping the car when I wanted to pick a roadside weed or offering encouragement for the crafts I made from sticks, leaves, bark, or other botanical remnants. I was a true nature nut. None of my local childhood friends shared this enthusiasm—and most definitely no one else in recent generations of the Lowman lineage had shared a passion for botany (although my grandfather obviously respected nature enough to retain an elm trunk amid his construction work). In rural upstate New York, we didn't have easy access to museums, scientists as role models, or other resources to nourish a child's love of natural sciences. All we had was outdoor play, but that simple pleasure transformed a small-town kid into a young naturalist.

Over long days outdoors, I developed patience for solo observations of the natural world that included many hours of silence. Such behavior may have reinforced my shyness. When I entered kindergarten, I became the class wallflower, miserable when confined indoors among noisy classmates. I almost never spoke in class except when called upon. The teacher told my mom something was quite wrong. I was carted off to our family doctor, who, in a gruff German accent, smiled and said, "Frau Lowman, have you considered the alternative?" On the last day of kindergarten, our teacher, Miss Jones, was grading workbooks. I admired her because she had humbly told us her story about why she had leg braces, living bravely with polio after some boys had pushed her into a stagnant pond during childhood, causing her to swallow filthy water. The teacher's story haunted me and reinforced my fear of bullies. That year, I had a perfect kindergarten workbook, until the final day when my best friend, Mimi, became jealous and circled the wrong answer on the last page using her big black crayon. With heartbreak, I dutifully handed it in without a word. Miss Jones sighed and said, "Oh, Meg, you almost had a perfect workbook until today's mistake." Tears welled, and I gazed in horror as Miss Jones gave me a silver (not gold) star for the year. I did not even have the courage to tattle on my best friend. For many years afterward, Mimi and I stashed the infamous workbook in our secret tree fort, giggling about its controversial history. (She is still one of my closest friends, despite the loss of that gold star!) I later lamented not having a bona fide naturalist or professional botanist in my childhood, to guide and inspire me.

After reading their biographies at the public library, Rachel Carson and Harriet Tubman became my role models. Carson confronted the chemical companies, having figured out that songbirds were dying due to pesticides. She quietly but forcefully told her story so the public could understand the science. Tubman led slaves north along the Underground Railroad at night using moss on tree trunks as her navigational guide—truly a pioneering naturalist. I used to close my eyes in the woods, feel for the moss, and try to find my way—it was never easy, so I admired Tubman even more. My only two role models were deceased—in hindsight, I guess that trees substituted as exemplary living entities and offered many life lessons. They stand tall, benevolently providing shelter, stabilizing both soil and water, and always giving back to their community.

My three best (and only) kindergarten friends lived nearby and played outdoors in the woods with me, although sometimes reluctantly. Looking back, I owe my scientific career not just to a passion for collections, but also to a few loyal compatriots who were willing to explore the backyard. Mimi was one of ten children in her family and my alter ego, because she was brave and outspoken. Betsy came from a family of nine children, where she had access to her older sisters' clothing; she was our fashion guru and much admired by all the boys. She was also my only friend who enjoyed birdwatching, which counted for a lot. Maxine was bold and funny, oftentimes blurting out crazy ideas. Once, she convinced us to smoke hollow sticks and we all thought our death from lung cancer was imminent. We were a devoted team, creating our little adventures before the town ever had cable television to watch National Geographic specials. A hundred feet behind my parents' house, we imagined it was halfway to Siberia on our "hullaballoo expeditions," a term we invented to secretly allude to our missions. Sometimes we took bologna sandwiches, a thermos of strawberry milk (my favorite), and a blanket to sit on. We always had some jars to collect critters, a plastic bag for plants, and a few empty shoeboxes for rescuing small creatures. Boys were not allowed. In those days, climate change was not part of the environmental vocabulary of youth, citizens, or even scientists, so the biggest threat to the local flora was gangs of teenagers running through the swamp and crushing the very blossoms I was so eager to collect. To avoid encounters with those boisterous boys, I learned to sit quietly and remain invisible in the woods, a good skill for a future field biologist.

The girls and I wanted a secret place where we could escape adults, boys, and all other distractions. We constructed a rough-hewn fort in the lower branches of a few birches and maples. Logs from my dad's woodpile and a nearby thicket of young saplings provided our initial building materials. It was by no means an architectural wonder, just a few footholds nailed into place, plus some branches and blankets hauled into a wonderful sugar maple crotch about four feet off the ground (although it seemed more like a great expanse when we were six years old). We ate lunch up there, drew pictures, and told stories in our special hideout. Despite its rickety design, we wiled away many hours caring for baby birds that had fallen from their nests, trying to repair butterflies with broken wings, or simply amassing the flowers I later pressed and stashed under my bed. One afternoon, we rescued earthworms cut in half by our dads' lawnmowers, trying to Band-Aid them back together, but the poor creatures did not survive our simplistic surgery. We role-played as explorers, nurses, heroines, scientists, and castaways. The birches, with their peeling white bark, inspired our imaginations to transform us into members of the local Cayuga tribe, who had used birchbark for canoes and other practical purposes. From that fort I learned the rudiments of forest succession, becoming aware of the tallest trees, strongest branches, shadiest canopies, and propensity of each species to house wildlife. Along with sumac and cottonwoods, birches were a relatively short-lived species in upstate New York, called early successional because they were the first to grow in a forest clearing, but their weak wood ultimately caused them to topple during high winds or snowstorms. Birches were then replaced by later successional (or climax) trees such as maple, beech, or hemlock. My parents had built our house on a cleared lot, so the backyard grew anew into forest; over the course of my childhood, several birches and cottonwoods that formed our playground later became overshadowed by maples in true forest succession. I watched cottonwood and birch transition into maple and beech, whose dense canopies in turn shaded out many of the wildflowers on the forest floor.

Throughout childhood, learning about the natural world—and especially all things floral—was my obsession. I became a local expert on phenology, the seasonality of nature's events, before anyone in Elmira, New York, had even heard the word. I knew exactly when and where in the forest to find jack-in-the-pulpits, followed weeks later by yellow trout lilies and amazing varieties of violets ranging in color from pink to purple to blue to white. Spring ephemerals are those early seasonal wildflowers that bloom before the trees leaf out, while sunlight still reaches the forest floor. This clever strategy allows them to grow and reproduce before the shady conditions from the canopy above prevent adequate light for flowering. Late spring and summer flora abound in sunny fields and open meadows, but not under heavily shaded maple or beech canopy. By the age of ten, I knew the calendar of phenological events for many wildflowers in upstate New York. I kept careful diaries to track all kinds of seasonality, from plants blooming to canopies greening to birds migrating to mosquitos biting to fireflies twinkling.

My wildflower collections grew enormous. I hoarded old telephone books under my bed, deploying them as plant presses, and checked out stacks of field guides from the public library to serve as identification aids. I am not even sure what inspired me to press my collections of flowers, having never seen an herbarium until college, nor did I ever meet a real botanist who would have taught me the technical nuances of how to collect plants. Somehow, I determined that a pressed wild-flower looked slightly better than a withered, dried-up carcass of stems after seeing many of my handfuls of wildflowers wilt pathetically on the kitchen table. Despite her patience toward my nature pursuits, Mom was unhappy about the mice attracted to all the pressed bits of roadside residue under the bed. She put out mousetraps with cheese, but thankfully those furry critters were well fed from the dried collections, so the traps never snapped in the middle of the night. I spent hours of most days squatting on the bedroom floor, which became my laboratory, poring over some crummy Golden Guides from the library, trying to identify specimens. When I opened the telephone-book pages after a month or so of pressing, there laid dozens of brown flat corpses. After all the effort of pressing and then waiting for the specimens to dry, it was disheartening to discover that most plants lose their color when dead. The challenges of coloration, plus a lack of any technical botany books, created extreme hurdles to identify many of the specimens.

My best reference for learning botanical jargon was a set of nature encyclopedias purchased at the grocery checkout display, which Mom kindly let me buy—one volume at a time, each for one dollar—when I helped her with shopping. I cherished all sixteen volumes, and they provided rudimentary definitions, including diagrams of pistils, stamens, and plant sex in a slightly more sophisticated fashion than the simple Golden Guides. I found it unsettling that the word "pistil," referring to a plant's male parts, sounded so like a deadly weapon, "pistol," but they were so radically different. There was so much to learn! I was just a small-town nature girl who loved the outdoors yet was not savvy about most technicalities of scientific vocabulary.

When our fifth-grade teacher casually announced the next New York State Science Fair would be held in the nearby town of Cortland, I was hesitant but also determined to enter my collection. Maybe the science fair would introduce me to other kids studying nature? I drew a poster illustrating the general parts of a wildflower: petals, sepals, pistils, and the rest. It was simplistic, but I felt as close to my hero Rachel Carson as ever, having cataloged several hundred types of local botany over the past five years as a "scientific collection," all carefully pressed and labeled in ridiculously small five-by-seven-inch photo albums. I selected four books of pressed specimens, which was about half of my homemade herbarium. Not really knowing that plants in a professional herbarium were glued on large eleven-by-seventeen-inch sheets of paper, I had purchased from the local drugstore commercial photo albums intended for baby pictures; instead they became laden with dried (and mostly brown) wildflowers and small index cards listing the name, date of collection, location, and habitat. I tried to pick the best ones, either those with a few vestiges of color or ones with really cool names (such as snakeroot, live-forever, or Indian pipe), and avoided showing the less attractive ones, such as a cattail that had virtually exploded with white seeds all over the page.

Dad, despite knowing absolutely nothing about botany beyond mowing a lawn and raking leaves, offered fatherly enthusiasm and awoke at 5:00 a.m. on the day of the science fair to drive me two hours to Cortland. I did not sleep a wink the night before, shivering in fear that someone might ask me questions about the project. Not only was my shyness magnified at the notion of such a public event, but I had never met a professional scientist. Carefully loading the albums of pressed plants plus the crayon posters into our secondhand 1953 Ford Crestline Sunliner, Dad and I set out on what felt like a great expedition. It was 1964, and he always bought gas when it was on sale, so it must have been a week of higher prices at the pump because he had neglected to fill the tank. Dad obviously did not want to worry me, so as we came over the crest of the hill into town, he said, "Hang on." The car glided downhill on empty, careening through red lights and quiet streets in those wee hours of the morning. We were first in line when the gas station opened.

The science fair was held in a huge gymnasium at the state college, and I was assigned a small table for display. Wedged in among what seemed like 499 unruly boys, I did not see any girls but hoped a few were scattered in the crowd. I yearned to find a kindred spirit. I was astonished at the multitude of tables demonstrating a chemistry activity to replicate a volcano: pour vinegar on baking soda in the middle of a papier-mâché pyramid and, voilà, an eruption! There may have been only fifty volcanoes in the auditorium of five hundred students, but they attracted raucous cheering and bawdy attention that only emboldened their creators and was simply not part of my DNA. Had I not felt so self-conscious, it might have struck me as funny—a wildflower collection displayed by a consummate wallflower amid the chaos of vinegar volcanoes. But I was overly nervous as the only botanist (and one of few females) in the auditorium, as the judges later informed me. Nor did I see any other natural history projects throughout the entire science fair. The judges passed by in a herd, without commentary except to glance through a few pages of dried flowers and offer polite comments about the challenges of pressing plants without harming them. (I wanted to blurt out, "Of course, you dummy, if you pick a plant, it most definitely damages and ultimately kills the flower.") Unlike most students, who came with classmates from their schools, I was the only kid from my elementary district, so not part of a gang roaming the aisles to gaze at other projects. I spent the entire day standing beside my wildflowers; even my otherwise loyal dad ran a few errands to while away the hours. After such a long day, I was anxious to pack up the display and head home to the sanctuary of my bedroom laboratory. Then, to my amazement, I was called onstage for second prize. Speechless, but feeling an unexpected sense of accomplishment, I received a small plastic trophy, and could only hope that Harriet Tubman and Rachel Carson were looking down from heaven in approval. In the eyes of my family, this award was akin to a Nobel Prize and resided on our kitchen table for months. Although it did not increase my popularity on the elementary school playground as would a sports award, it gave my parents a glimmer of hope that their daughter's unusual love of nature might yet reap rewards.

After conquering roadside botany at the fifth-grade science fair, I stumbled by accident into an ornithology project a few years later. While cleaning our grandparents' attic, I found two dusty old wooden cases of birds' eggs collected by a nineteenth-century ancestor. (Maybe a family nature lover whom no one had ever talked about?) These exquisite ovals were swirled with blue, gray, white, and cinnamon by

Mother Nature's paintbrush. But the labels had disintegrated, victims of book lice. My grandmother was a formidable English professor and impeccable housekeeper, and doubtless considered these eggs a disgusting mess. So she allowed me to haul the treasures home to my bedroom-floor laboratory to attempt to identify them. Taking another big trip to the library, I exchanged botany field guides for bird books. It was relatively easy to find volumes that illustrated the birds themselves, but much tougher to sleuth out egg descriptions. I soon learned coloration was a critical part of classification, along with shape and exact size. No ordinary ruler would suffice; bird eggs needed more sophisticated instrumentation. A few of the library books mentioned specialized calipers to measure eggshell thickness and dimensions. I courageously wrote a biological supply company whose ad I saw in an *Audubon* magazine and requested a catalog. I had a small allowance from my household chores and persuaded my mom to write a check for this mail order if I repaid her from my piggy bank. For only $13.95, I soon owned a set of calipers, and spent hundreds of hours measuring the egg dimensions of the wood thrush, Baltimore oriole, robin, goldfinch, killdeer, and more.

Bird egg identification was trickier than botany, almost impossible with rudimentary field guides. The descriptions in generic bird books were usually limited to "medium-sized, blue" or "solitary white egg." I dug deeper into the ornithological literature and pulled out dusty volumes at the public library by John Burroughs, John James Audubon, and other nineteenth-century naturalists. Eventually, I taught myself a new vocabulary: metric (not inches); patterns such as streaked, spotted, splotched, or speckled; and how to differentiate between cinnamon, hazel, chestnut, and brown. I didn't have an approachable science teacher, or even one friend at school who shared a similar passion for birds, so this was a lonely pursuit. At the time, out of about one hundred members of our local Audubon Society, I was almost the only one under the age of seventy and could not envision any of them squatting on the floor to gaze at bird eggs. My parents had given me an Audubon membership as a birthday gift, and I eagerly attended all their nature films shown in a local auditorium.

The senior birders sometimes kindly invited me on a Saturday outing. They more or less adopted me, explaining how to estimate bird counts of a migratory flock and identify confusing fall warblers, one of the biggest challenges for any amateur bird lover. Although a neophyte, I was truly captivated by birdwatching, but too tongue-tied to tell them about my egg collection.

Occasionally, Mom would drive me to the Cornell Laboratory of Ornithology to walk the trails and watch birds. Just an hour away from home, the lab included a public display with bird sounds piped from a pond into a room with a big window. It was a thrilling sensation to hear the Canada geese's honks or the mallard ducks' squawks or the spring songs of returning migrants. On two occasions, I also stood shyly for at least an hour in the hallway outside the scientists' offices, hoping someone might talk to a small girl holding a Tupperware container of bird eggs. No one ever did, despite those hopeful attempts. Both times, I brought a few problem eggs, hoping to meet an expert who would quickly resolve their correct identification. It would have been beyond wonderful to talk to a bird scientist. Recalling that enormous disappointment when no one took notice, I now respond to every kid who contacts me, without exception.

At some point, I became stumped by a large white egg. Almost a year later, I returned to the mystery of that plain white egg once again. I compared it against all the photo guides, measured and remeasured its length and width, and puzzled for weeks. It was larger than warbler or thrush eggs, and it also lacked any unique coloration. Then one Saturday, while scrambling eggs for breakfast, I had a eureka moment and realized the answer had been in front of me all along! I grabbed an eggshell from our kitchen and ran upstairs. Wow! It was almost identical to the mystery egg. I had spent a year imagining this was the egg of a great auk or a whooping crane, yet it turned out to be an ordinary chicken's. Having just reread *Silent Spring* by my hero Rachel Carson, who explained pesticides were killing songbirds, I was inspired to design a remarkably simple science project. During the twentieth century, Carson discovered that birds (even chickens) ingested pesticides, and the toxicity resulted in eggs with thinner shells that never hatched. My dusty old eggs had been collected in the mid-1800s, making them approximately one hundred years old. I took several eggs from Mom's refrigerator, dated 1970, and measured their thickness with the calipers. Then I delicately cracked the ancestral egg in half, calculated the shell thickness, and compared it to the modern eggs. The hundred-year-old eggshell was 0.019 inches thick, while a recent eggshell averaged 0.011 inches thick. (My calipers used good old-fashioned inches, not centimeters.) I later learned about statistics, and how scientific research with only one replicate is not very robust, but at the time it seemed like a breakthrough. I housed this mini research project comparing eggshells one hundred years apart in a modest cigar box with handwritten labels summarizing its findings, and it continues to merit a place of pride on my library shelf, reminding me about the thrill of scientific discovery.

By junior high school, my three-musketeer girlfriends had substituted boys for tree forts. But my passion for nature grew, and I eagerly spent weekends birdwatching in the local parks on Saturday. As an obsessive list-maker, I recorded all sightings: evening grosbeaks were a highlight, as were woodpeckers with their incessant drumming on all the different dead trees. Growing up in small-town America was a mixed blessing. We knew the soda jerk. We walked to school, played outside, shoveled snow, picked blackberries in fields, and caught fireflies. But the public school had its share of bullies and substance abuse, so finding friends with nerdy interests like birdwatching was difficult. I was so determined to find a friend who liked nature that I wrote to Duryea Morton, a prominent leader of the National Audubon Society whose name I found in the magazine. I explained to him I was a birdwatcher, and did he have any advice to find kids who shared my passion? Miraculously, Morton wrote back from his lofty office in New York City and offered a solution. He suggested attending a summer camp in West Virginia run by an ornithologist friend of his named John Trott. It was the only nature camp for youth in all of America at the time, and he thought I just might find some fellow bird lovers among the campers. Although my parents were not keen on driving me as far away as West Virginia, and camp tuition was quite a financial stretch, they reluctantly signed me up the following summer for Burgundy Wildlife Camp, fervently hoping I would no longer be relegated to Elmira's seventy-plus-aged birders for social activity. After an all-day drive from Elmira, New York, to Capon Bridge, West Virginia, the final dirt road into camp curved around a still, with a colorful crowd of locals enjoying the moonshine of their labor! We forded a stream to cross into camp property, and it was a miracle Dad didn't turn around. The camp was rustic and gloriously situated in the middle of the woods—a creek for catching aquatic critters, canopies bursting with songbirds, miles of hiking trails, nets set up for bird banding, a fireplace surrounded by screened porches with all sorts of wildlife-collecting gadgetry, and, best of all, nineteen other kids who loved nature.

During two life-changing weeks of camp, I found friends who paid attention to ants, rocks, wildflowers, salamanders, mosses, and, yes, birds! It was truly heaven on earth, and the directors, John and Lee Trott, became lifelong mentors for me and many other campers. I was assigned a top bunk in the girls' dormitory, a simple open-air construction composed of screens and rough-hewn logs that accommodated a dozen girls. The first night, I felt a whooshing just inches above my face and huddled under the blankets in fear, wondering what was attacking me. The next day, when I mentioned this assailant, the camp counselor explained that a bat lived in the rafters just over my bunk, and how fortunate for me because it would eat all the mosquitos. Her explanation alleviated my panic, but just a little. All day long, I quietly digested her statement and ultimately became convinced a resident bat was indeed an exceptional roommate.

Wildlife camp changed my views of the natural world in many ways, not just about bats. I held a live goldfinch in the bird-banding tent, an extraordinary spiritual experience for a bird lover like me. I learned the constellations under a night sky in the wilds of West Virginia with total absence of artificial lights. We swam in a muddy pond as our only sporting activity, and when I found a pollywog flattened under my bathing suit, the camp director congratulated me for attracting one of the water's precious bits of biodiversity. I was still shy and worked extra hard to pull the bedsheets perfectly smooth so as not to be called upon during daily inspection, when the dormitories were checked for neatness. I desperately wanted acceptance into this nature gang. At camp, all kids were considered naturalists regardless of background or gender, so I developed close ties with boys (as well as girls) for the first time in my life. Most became lifelong friends, many became natural science professionals, and we continue to go birding or botanizing in our adulthood.

Every camper undertook a research project, so I decided to identify the mosses of a West Virginia forest. Having tackled wildflowers and trees in upstate New York, I was ready for the challenges of nonflowering plants. And I also wanted an excuse to use the amazing camp microscope that was required to identify mosses. Harriet Tubman had mastered the mosses of the forest to navigate the Underground Railroad, and I was determined to walk in her footsteps as the camp bryologist, aka moss expert. My enthusiasm was so great for these smaller bits of fuzzy green stuff that after I'd created a detailed moss collection, the Trotts asked me to return the following summer as a staff member. A thirteen-year-old joining the staff? I was humbled. The directors fervently believed "children educating children" was the most effective model for learning, and they hired teenagers to teach the campers. The paycheck for that first summer job was a whopping $25, but I felt very wealthy. I could earn the same amount after one long night of babysitting back in Elmira, but camp paid me in other indirect ways. Thanks to one bold letter sent to the National Audubon Society, I now had a cadre of friends who loved to talk about bird migration and tree identification, plus I learned some skills for teaching younger students about the natural world.

My first teaching assignment as a teenage camp counselor was trees (dendrology). Not surprisingly, I was tongue-tied and inexperienced, but the camp director assuaged any anxieties by reminding me that learning with your students is a more effective teaching mechanism than preaching facts to any classroom. Nonetheless, I was extremely nervous. During the preceding winter, I'd signed out a big pile of tree books from the public library and eagerly read and reread every page.

FROM WILDFLOWER TO WALLFLOWER · 23

With index cards in my pocket, I returned to camp the following summer as the official dendrology teacher. I gave lectures outdoors under a glorious red oak, engaging campers to become tree detectives by collecting acorns, measuring trunk girth, and examining leaves for signs of insect attack. Both a love for trees and encouragement from the camp directors transformed me into an enthusiastic educator. I worked at Burgundy Wildlife Camp for six summers, teaching about spiders (arachnology), insects (entomology), and geology. Thirty years later, I returned to the camp and built a canopy walkway in that same magnificent oak, so campers have the privilege of exploring its aerial secrets. Now the next generation of campers at my childhood summer camp are arbornauts!

Back home during high school, I had plenty of fields and woods to explore, but no Smithsonian Institution to visit on the weekends (as did most camp friends who lived near Washington, DC) and no student internships in nearby technology headquarters or environmental organizations. In Elmira, students frequently hosted beer kegs on the hill behind our school, smoked in the parking lot, bragged about bad grades, or hung out in the basement of a rundown store called the People's Place, which sold bell-bottom jeans and was run by a wonderfully rebellious classmate by the name of Tommy Hilfiger.

A gawky seedling struggling in poor soil, I realized that plants were a lot like me because they didn't talk. A gregarious toddler is probably attracted to playful puppies, and maybe creates volcanoes for a science experiment, but I was smitten with wildflowers and studied all their botanical parts including the pistils. Not the kind that shoot bullets, spelled "pistols," but the innards of flowers where all the sex occurs. Instead of collecting Beatles records, I collected beetles with six legs. Instead of doing my nails with pink polish or discussing new hairstyles at sleepovers, I asked if anyone wanted to wake up early for birdwatching. It was considered "cool" to skip school and not to make the honor roll. Those were the dilemmas confronting public school students in a community that lacked a strong economic base during the 1960s. Upstate New York was part of a growing American malaise of unemployment and food stamps, unlike cities where innovation and technology led to vibrant economies. Growing up in America is a lottery, where your zip code can often foretell your future.

When I decided to apply to Williams College, the high school guidance counselor told me there was no such place—surely, I meant the College of William and Mary. I knew better because I had bought some college catalogs and read that Williams College in Massachusetts was one of a handful of small schools with its own forest. At the college interview, I was a nervous wreck, shaking like an aspen leaf. The admissions officer noted from my application that I had taught spiders at summer nature camp. He looked up at me very seriously and asked, "Margaret, exactly what did you teach the spiders?" It seemed obvious he was not making a joke. Horrified I had not made the college essay completely clear, I quickly explained I taught the subject of spiders to kids, not the spiders themselves. This miscommunication convinced me that admittance to Williams College was squashed. Several months later, to my total amazement, I received an acceptance letter.

When the high school principal announced I was not the class valedictorian, but came in second as salutatorian, my fellow classmates were upset—the valedictorian had not taken an honors courseload, which made her less deserving in their eyes. Much to my surprise, the class voted for me to give our graduation speech. Suddenly those years of academic dedication bore fruit. Graduation was a big deal in a small town, and everyone was counting the days; in contrast, the prospect of giving a speech to an entire auditorium of people made me lose weeks of sleep. I rehearsed in the shower, woke up in cold sweats while reciting in my dreams, and all but had a nervous breakdown in anticipation of graduation night. But then it rained . . . and rained and rained. The big flood of 1972 is etched in the history of the Susquehanna River Valley; Elmira's local Chemung River was a tributary. The river broke its banks around 2:00 a.m. on June 23, and many high school seniors awoke during the last week of school to find several feet of water in their kitchens and living rooms. Families evacuated. Schools closed. Roads flooded. The high school became a Red Cross emergency center. Dead bodies were fished out of surging river waters.

The smell of mud oozing in homes and mold in walls was a stench I will never forget. A town with a bleak economic outlook became even bleaker. Suddenly my classmates and I were volunteering to give tetanus injections instead of dressing for a prom. What a bittersweet ending to our high school years! We never officially graduated but received our diplomas in the mail a few months later. Elmira and her surrounds never really recovered from those floods. The real-estate values plummeted, especially homes close to the Chemung River. As families moved away, school enrollment shrank; our former high school no longer exists. Dad's bank downsized and he lost his job. The flood was a nail in the coffin of an already fragile economy. Mother Nature rules without exceptions. At that time, extreme weather was considered a hundred-year event, so we considered the flood of 1972 an anomaly. A mere twenty years later, the rapid onset of climate change would make flood frequency—as well as droughts, fires, and heat waves—a commonplace occurrence in many regions. Our Elmira flood was a harbinger of things to come.

Much has changed over the past fifty years, not just climate change and the frequency of floods and fires, but also the science of plants. There have been many advances in the ways we collect and preserve plants, the ways we identify them, and the manipulations agricultural scientists execute to create hardier crops or disease-resistant elms. But two of my biggest life lessons as a scientist started with those childhood plant collections in rural upstate New York: (1) "the power of one"—by making observations, usually solo, I not only mastered local wildflowers but also become an amateur expert on birds eggs, all baby steps toward becoming a professional field biologist—and (2) "start local, but go global"—a personal reflection looking back because by learning the landscape of my backyard and then later ramping up to global ecosystems, I became a more effective field biologist. Thirty years later, my childhood tree fort transitioned into tropical canopy walkways on several continents. A love of one tall elm in our lake cabin expanded to an international forest conservation profession. The joy of playing outdoors throughout childhood—cultivating my five senses to find, touch, smell, and identify plants—inspired many subsequent years of undergraduate and graduate student training, not only my own, but also mentoring women like myself and minorities. All of these childhood passions, patched together like a quilt, led me to ultimately become one of the world's first arbornauts. I probably would not have pursued field biology as a career without a halcyon childhood of outdoor exploration. Mostly trees. Mostly solitude. Mostly wildflowers, leaves, and a curiosity about how nature operates.

»» American Elm ««
(*Ulmus americana*)

AMERICAN ELM (*Ulmus americana*) was first classified by Carl Linnaeus in 1753, when he wrote the famous *Species Plantarum*, which paved the way for all future classification of organisms. This tree is a member of the family Ulmaceae with six genera and forty species. It is native to North America, and has a distribution along the east coast from Maine to Florida, west to North Dakota, and down to Texas. Its British cousin, English elm (*Ulmus procera*), was so common it had a nickname, "Wiltshire weed," because its silhouette dominated the landscape. American elm was similarly widespread, and grew along floodplains, stream banks, swampy terrain, hillsides, and well-drained soils. In short, it grew almost anywhere! Because the seeds are wind-dispersed, elms spread rapidly and germinated almost immediately. Elm bark was used by Native Americans for various medicinal applications, and the wood was coveted for furniture, flooring, caskets, and crates. The elm canopy used to represent a lively epicenter for birds and butterflies, as well as a host of herbivores including leaf miners, borers, mealybugs, and scales. Also called white elm, American elm is the state tree of Massachusetts and North Dakota. Historically, it was fast-growing, hardy, and successful in urban settings, making it an important street tree throughout its range. A relative of the American elm, the slippery elm (*Ulmus rubra*), was commonly sought to soothe the digestive tract, and its rough leaf surfaces were used as a natural rouge for women in colonial times: a brisk rub with slippery elm foliage left the cheeks irritated, red, and allegedly more beautiful. As kids, my girlfriends and I loved to scrub our cheeks with slippery elm and pretend we had applied makeup.

Elms were the most common species growing in New England villages and towns at the turn of the twentieth century, although no satellite imagery existed in those days to provide exact counts. During the mid-twentieth century, over 99 percent of American elms were killed by Dutch elm disease, a fungal infection introduced from Europe that created die-offs in the Northeast and throughout the species' entire range. The only survivors were isolated individuals, notably in Florida and British Columbia where the disease had not spread outside of elms' normal distribution. Dutch elm disease not only affected American elm but also killed off English elm in Europe during the 1970s. The fungus *Ophiostoma novo-ulmi* (also known as *Ceratocystis ulmi*) was spread by bark beetles of the genus *Scolytus* (and later another bark beetle genus called *Hylurgopinus*). Female beetles sought weakened elm trunks to excavate their egg-laying galleries between the bark and the wood. If the fungus was present on the female, numerous fungal spores were deposited inside the egg chambers. When young adult beetles emerged, they then carried and distributed the spores to healthy elms as they flew off and fed on their foliage. Spores infected the xylem and reproduced inside these water-conducting vessels almost like a fermenting yeast. As a newly infected elm weakened, it also became susceptible to beetles seeking out dying trunks to lay eggs. The elimination of Dutch elm disease required complete extermination of beetles, which was expensive and almost impossible under natural conditions.

Throughout the twentieth century, American elm was characterized by geneticists as tetraploid, meaning it had double the number of chromosomes of its counterparts classified as diploid. However, more recent genetic analyses now reveal some elms are diploid, and those individuals seem more resistant to Dutch elm disease. Having two subspecies of elms might lead to developing genetically resistant elms for future propagation. Hybrids of American and Asiatic elms are also showing increased resistance to the fungus, offering hope for future restoration of elm canopies.

Elms probably inspired my career in botany, thanks to my grandfather's clever construction of our lake cabin. I went from a childhood tree-love to study the phenology of temperate forests as part of an undergraduate biology thesis. During those years, I also suffered emotional pangs when the elms on campus were cut. Dutch elm disease was an enormous setback not just for forests but also for urban regions, where canopy shade enhances real estate values and provides enormous ecological services, otherwise known as "natural capital." In addition to shade, urban trees provide water filtration, soil conservation, and carbon storage; cleanse air pollution; and offer a home for many birds and animals. Trees supply ecological services worth billions of dollars if we only allow them to grow. In my hometown of Elmira, New York, enormous elms lined the city streets until they were ravaged by Dutch elm disease. I really miss them.

2

BECOMING A FOREST DETECTIVE

֍֎

First Encounters with Temperate Trees from New England to Scotland

WITHIN THE FIRST THREE WEEKS of college, I found out
that 95 percent of biology majors at my school were premedical and most of the coursework focused on blood
cells, not bird songs. Medicine was not my intended career track so
I searched the catalog for an alternate major that might offer more
outdoor focus. I turned to geology, thinking, Why not study the
landscape where my beloved forests grow? This seemed like a good
plan B to achieve my career goal of working in nature. I spent most
of those first-year geology field trips photographing wildflowers and
birds while everyone else focused on bedrock, but I was thrilled to be
accepted into a joint summer field course with the University of Idaho
and Williams College geology departments. My enthusiasm waned
to discover the ratio of one female: nineteen males. But I wasn't going
to miss this opportunity, so I packed hiking boots, blue jeans, sweat-
shirts, and binoculars, eager to spend a summer learning about the
magnificent Rocky Mountains. I'd never traveled west of the Mis-
sissippi River, and I immediately fell in love with the landscape, happily drawing rock formations and imagining dinosaurs standing amid
these giant geologic outcrops. During our very first week clambering

over metamorphic ridges, however, I watched helplessly as most of my classmates spent their lunch break throwing their rock hammers at ptarmigans. These geology hammers had a very sharp claw, intended to split open extremely hard basalt; the boys thought it was fun to heave them at these gorgeous ground birds. Fortunately, their aim was not great, probably because the hammer was not weighted to act like a javelin. Yet the meanness of their behavior and the fact that, despite my passion for birds, I didn't dare speak up to these hulking "jocks for rocks" left me feeling disgusted with them and disappointed in myself.

I spent my free time jogging, partially to escape the social interactions at base camp but also because the grandeur of the surrounding tall peaks was breathtaking. The extra exercise on top of an arduous day of outcrop mapping really tired me out, so I crawled into bed early while the rest enjoyed beer and poker into the wee hours of the morning. One day, while running on a lonely highway, a pickup truck drove so close that the cowboy in the passenger seat smacked me on the butt and made some lewd comments. After that, I stopped running on roads and instead sprinted up mountain trails, risky due to the uneven rocky terrain, but it felt safer than confronting highway miscreants.

That summer experience in Idaho really tested me, because it became obvious that geology was a white-male bastion, at least back then. I did not feel welcome, nor did I seem to belong in a community where excessive beer drinking, attempts to assassinate ptarmigans, and rough language ruled. When I returned to classes the next fall, the head of the geology department seemed surprised I had finished the course, much less earned a good grade. He made it clear a female student had never enrolled in his senior capstone class, which was required to complete the major. It was not an outright rejection, but his negative, guarded warning made clear I would not be welcome. As the only aspiring female geology major at that time, I did not have any peers with whom to join forces or commiserate, nor did it dawn on me to report this experience to the administration. Those were the early days of coeducation for many colleges; I was part of only the second class to include women at Williams. So I simply internalized the disappointment and sought an alternate path. With almost a sense of relief, I changed my major back to biology. After all, I had originally applied to Williams because of its forest.

In addition to the challenges of the gender ratio in geology, I almost flunked mineralogy—somehow, public high school had not prepared me for the nuances of college study habits and essay exams. I buckled down and practically memorized the textbook. With only a C in that class, I increasingly had doubts about a science career. I thought about those days of standing outside the office doors of ornithology professors at Cornell University and realized I had never met a living female scientist (even at Williams, the science faculty was all male). The biology textbooks, as well as the geology books, consistently featured male role models. It gave me a sinking feeling . . . was there truly any place for women in field biology?

Starting up my biology major as a rising junior, I needed to design a thesis in a hurry if I intended to pursue field research, which, by definition, requires repeated years of seasonal data collection to average out anomalies like droughts or extreme winters. So I went back to my roots and expanded those childhood observations of wildflowers into a proposal to study the phenology of the biggest plants: trees. I spent the final two years of college camping out in the forest, immersed in the joys of nature, and surrounded by my old friends: maple, birch, oak, and beech! Only one professor taught any natural history courses, so it was a no-brainer to request him as an advisor. Given the predominance of the premedical students, he seemed enthusiastic to have a budding plant ecologist under his wing. I was eager to collect real data using a scientific protocol instead of just amateur journaling. I had some exposure to professional research from the readings in my ecology course, and quickly learned about the methodology for designing fieldwork: ask a question, locate a study site, collect data, research the existing literature to find similar work, analyze the results, and then write a paper. I had grown up watching leaves regularly emerge each spring and fall every autumn, like clockwork, and wondered if the wood growth of temperate trees was as seasonal as the leafing. At the outset, it seemed easy to design a field research project, almost like following a recipe in a kitchen. I selected sixteen tree species and formulated a hypothesis that their trunks expanded every spring and then shut down in the fall, just like their seasonal foliage. Unwittingly, I followed the conventional footsteps of two hundred years of foresters, limiting my view to a tree's "big toe" at ground level instead of thinking about the whole forest. At that time, a ground-based perspective was also the easiest way for scientists to view a tree, focusing on its trunk. Those tall brown pillars that support the foliage high above are composed of 99 percent wood (dead cells providing mechanical support) and approximately 1 percent living cells in a narrow band called vascular tissue that grows just under the bark. The vascular tissue consists of two skinny, one-cell-thick highways: xylem, which carries water up from roots to leaves, and phloem, which brings sugars down from the chloroplasts in leaves to the root hairs below. The only living section of a trunk, vascular tissue grows and dies back every year, a seasonal process resulting in annual rings, which scientists now consider the bar code of climate history. Over the long haul, tree rings tell stories not just about a tree's life, but about the environment around it. Tree ring experts, called dendrochronologists, can track Earth's environmental history, including episodes of drought, fire, and good growing seasons. As a budding field ecologist, I wanted to answer questions about the short-term dynamics of trunk growth over months and seasons, not the entire lifetime of a tree. Eager to embark on my first official botanical field research, I felt like a racehorse ready to break out of its stall.

I enthusiastically read existing forestry publications about temperate tree trunk growth. There was not too much ecological literature, but plenty about maximizing tree growth in pine plantations, since timber production was the most heavily funded aspect of tree science at the time. It was no surprise that most research on trees focused on timber and the economy of board feet, which is a standard unit of measure for lumber. The college biology library had extensive shelves of scientific journals, a year's worth of each publication bound in colorful leather ranging from red to bright orange to fuchsia. I combed the dusty shelves over many nocturnal visits since the stacks were open twenty-four hours a day. I became a regular along with all the premedical students, who seemed to live full-time in the science library. My field biology journals, including titles such as *Botanical Gazette*, *Plant Physiology*, and *Botanical Review*, were extremely underused compared to titles like *Cell Physiology* or *Blood and Cells*. I probably broke the stiff bindings of every single issue of tree-related journals from about 1950 to 1974. For pre-1950 publications, students submitted a written request for archived journals, which were then brought out of storage off campus. This loan process usually took a week or two. No digital copies of scientific journals existed, just enormous bound volumes.

For the first time in my life, I would be collecting real data. This required some specialized equipment. Studying the methods used by foresters, I came across a very cool gadget called a dial-gauge dendrometer, which measures trunk growth, also known as girth expansion. It was not too expensive, so I ordered two (best to have a backup in case one breaks) and then set about gently affixing a template of screws at breast height into the trunks of sixteen species. "Breast height" is a rough forestry term, defined so different girths can be compared at a similar height above ground. (Of course, my breast height was undoubtedly lower than the average logger, ever envisioned by the public as someone resembling Paul Bunyan.) The four screws on each tree created a square array or template upon which to rest the measuring dial-gauge that estimated the distance from the screw heads down to the bark, so that I could record trunk expansion, i.e., how much the bark had lifted up around the screws. As with hanging a hammock, small screws inserted into living trunks are not harmful, because 99 percent of all wood underneath the bark is dead woody cells, also called heartwood. (Besides cutting the entire tree down, the only way to damage a trunk is to girdle it, strangling its thin layer of vascular tissue located just under the outer bark, which would prevent water from moving upward from the roots and sugars downward from the leaves.) The dendrometer measures to a 0.001 inch, a very tiny increment of expansion, sensitive enough for a tree's extremely slow growth. I already knew from my childhood observations that oak leaves emerged earlier than beech and maple, but would their woody trunks be equally responsive to the earliest signs of spring? I also learned from reading all those technical forestry journals that trunks expand and contract with environmental conditions in both the short term (months) as well as the long term (years), so I would need to measure for more than one year to even come close to determining any patterns of trunk expansion. My awesome dendrometer was so sensitive it not only measured expansion but also contraction. In the case of drought or aging, a tree's girth reduces, like humans who lose weight when starving or height when they grow older.

I only had two years to measure, measure, measure. And so I did—sixteen trees measured twice a week for twenty-four months, in snow and ice, in rain and lightning, as well as in heat waves. (I had two gaps in my data for the weeks of Thanksgiving and Christmas.) Most of my premed biology classmates were safely confined to a laboratory, with comfy couches, soda machines, and pizza on order, conducting thesis experiments on mice or cells. But I spent my days in the college forest, crouching beneath seventy-five-foot trees, portioning out my quart-sized canteen of water, munching trail mix while I took 736 measurements (to be exact) of tree trunks over two years and documented observations of their leafing phenology. During winter, my fingers froze as I wrote copious notes in my field journal, and my feet tingled from numbness while standing in snowbanks at the base of each tree. Hopkins Memorial Forest, the field property of Williams College, became my second home. I rode a bicycle the five uphill miles to the field site and trampled a path to and from each of the sixteen trees. Sometimes I brought back large branches of spring foliage to measure bud or leaf size, causing my roommates to chuckle about all the vegetation in our dormitory living room. Word got out that I was an expert on the college forest, and it became a social event for students to accompany their resident nature nut to measure her trees. Even a few premed students ventured into the woods and found it charming to learn bird songs or identify spring wildflowers. We traveled the forest on skis, on snowshoes, on foot, and even barefoot. One friend had an alleged passion for botany (but the kind that could be smoked), and he repeatedly begged me to awaken him at 6:00 a.m. for a nature foray. I would knock, and he would stagger to the door with extremely bloodshot eyes, laugh out loud at the sight of me dressed in khakis with binoculars in hand, and collapse back into bed.

During the two years of my thesis research, I never focused on the treetops. So like many generations of foresters before me, I overlooked about 95 percent of the forest. I got excited about wood—not board feet, but how much grew in a healthy temperate forest and how the seasonal growth patterns varied between neighboring species. Two physiological types of wood growth predominated in the existing forestry literature, each producing different patterns of annual rings. One was called diffuse porous, where species such as maple, birch, and beech lay down tree rings of uniform-sized woody cells. The other was ring porous, where large woody cells were laid down in early spring and smaller ones later in the growing season, as exemplified by elm, ash, and locust. When I looked at the ecological role of diffuse-porous species, I realized they were also late successional, meaning they were winners in long-term growth competition to achieve canopy status. The ring-porous species, conversely, tended to be early successional: they grew first and fast on a cleared hillside, but ultimately lost out to later successional species that grew slow and steady, overtopping them. The correlation between wood growth patterns and successional status of a tree species was a new discovery, at least to me. I had not run across this information in any of the dusty forestry journals available at the college biology library and couldn't be sure if it existed elsewhere, but I fervently hoped it was a new finding. My advisor agreed it was a promising lead. This finding led me to wonder if the patterns of leafing and wood growth were similar. Did trees that leafed out first also have early spring wood growth spurts? During year two, I paid close attention to the leafing patterns of each tree as I measured trunk growth. Did early trunk expansion allow some species to outcompete others through an early growth spurt each spring; or alternatively, did late, slow-and-steady summer wood growth enable tree species to ultimately dominate the canopy? That type of long-term research was beyond the scope of my thesis, but it certainly illustrated the importance of data collection over the lifetime of an organism, which, in the case of trees, is a long time! Suddenly, I found myself asking one question but raising many more. Eureka! Little did I know this was what made up the work of a scientist and what kept it so exciting.

I measured trees with unflagging devotion, and in exchange, they shared some of their secrets with me. I learned that girth shrank during sultry summer weeks. Both sides of each trunk consistently showed the same seasonal growth responses, although each side fluctuated daily—slight shrinkage when sunlight hit at different times of day as well as micro-expansion with rainfall. My most exciting discovery was undoubtedly the seasonal patterns of trunk expansion between those physiological growth categories of ring- and diffuse-porous species. Ring-porous trees produced an early spring burst of trunk expansion and late leafing; diffuse-porous trees, with wood growth evenly distributed over the entire growing season, leafed early. My first-ever research hypothesis, that trunk growth exhibited seasonal patterns, was confirmed. But the real discovery came from poring over the data after finishing all my fieldwork. This led me to determine that diffuse-porous species (with their early budburst and evenly distributed wood growth over the entire summer) were also late successional species, the ultimate climax trees. In contrast, ring-porous species, which prioritize early wood growth but leaf later, were early successional; they grew quickly in a young forest but lost out in the long-term successional race for canopy dominance. Again, my results led to more questions than answers. Question one: How many samples is enough? In other words, what inferences could I draw by measuring only two sides of one trunk per species? My data were not at all conclusive but served to inspire future research. The challenges of sampling made me uneasy, even before I took a good biological statistics class. Question two: How might longer-term bouts of climate affect tree stands over their lifetime, when I was merely studying them for two years? I saw the immediate impact of weather, where summer rain caused trunk expansion and occasional dry spells shrank their diameter, but I would need several decades (or more) to tease apart the complexities of climatic fluctuations on forest health. This question bubbled up as I spent many hours walking in the woods in solitude. The significance of climate change had not entered ecological discussions, except in geochemical and climatological circles, so I had little context. And question three: Was it appropriate to extrapolate from one college forest tract to the entire Eastern Seaboard of temperate woodlands? I knew that similar stands of oak, beech, maple, and hickory existed not only in Massachusetts but also throughout portions of six other New England states. So it seemed logical that similar growth patterns existed elsewhere throughout the range of these tree species. But it also made me curious how temperatures (such as cold exposed hilltops in northern Vermont versus warmer valleys in southern Connecticut) might impact the phenological patterns of trunk growth and leafing. And what about big droughts? Or early snowfalls? All of my questions required more years of sampling beyond the scope of an undergraduate thesis. And clearly, I needed to learn more about statistical sampling to design appropriate field research.

Over those two years, I had to find a way to stick around campus all summer to continue monthly measurements of trunks when classes were not in session. I took on multiple jobs, ranging from dishwasher to payroll clerk to bartender to babysitter. The most relevant job was when my thesis advisor hired a small team of students to calculate how much wood grew in New England, as part of a study measuring potential fuel for wood-burning stoves during winter. Five of us spent one summer using a chain saw, which filled me with guilt but also provided important insights into the "big toes" of temperate trees. The project required us to cut down twenty-four individuals, weigh their trunk biomass, then estimate wood per acre of forest. Ring by ring, the chain saw ripped through an entire lifetime of each woody denizen we harvested. It was bittersweet, watching layers of xylem and phloem disintegrate into sawdust. One of our team was a forestry trainee from a nearby vocational college; he loved the chain-saw work, and zealously oiled and sharpened our toothed monster. It was helpful to learn about his perspective, which was entirely board feet and timber productivity for profit; through that lumber lens, he had a great enthusiasm for trees that I truly admired. I was probably more of a tree hugger than chopper, so the economic issues were not foremost in my mind. Years later, I was grateful for exposure to his more applied perspective on the value of timber since it remains an important element of forest management. Although we successfully created the energy model as assigned, the experience convinced me I wanted to study living forests, not forest products. It also promoted lively discussion, because wood-burning stoves, initially considered a boon for energy independence in parts of North America, soon after became a curse. Today, scientists consider black soot one of the biggest threats to climate change. When wood burns, it emits soot, and those particles not only lead to respiratory ailments for women and children who usually hover near cookstoves, but the soot also lands on winter landscapes where its dark coloration causes faster melting of glaciers and snowpack.

In between measuring trees, my last college summer was spent working as a research assistant for the college biology department, which involved sorting through a dusty and long-ignored herbarium. It was housed in the basement of the biology building and consisted of several thousand dried specimens. I loved poring over their brown, dead carcasses, a reminder of my childhood bedroom-floor laboratory. The conditions were slightly improved—I had graduated to a card table with a tiny white table lamp in a dark room. But I soon learned the environment was significantly high-risk because this space was shared with the departmental taxidermist, whose behavior was more than extremely questionable. Hired by the college decades earlier—back in the days when students studied organisms more than cells—he lurked in the basement, stuffing specimens. During the semester, biology students who wanted a daring midnight study break from the library explored the cavernous basement of the building, where hundreds of dirty stuffed animals were piled in dark rooms. None of us knew from our nocturnal explorations that a living human curated the dusty collections by day. It turned out the taxidermist was a predator himself—on me. Cataloging dried plants with his heavy breathing and wandering hands creeping up behind me was not a workable situation. After several awkward confrontations, I gathered the gumption to report him to the biology chair, who quickly "promoted" me to a desk upstairs. In addition to organizing the dried plants for the college, I had time to collect field measurements all summer, completing a full twenty-four months of trunk growth dynamics. My advisor promised to shepherd me through a first scientific publication, because he thought the data were original. I was beyond ecstatic, and duly wrote my share just after graduation. Then, I waited six months and wrote him again about our joint publication. I guess he got busy and forgot, because I never heard back. Later, at graduate school, I was envious of other students who had copublished a first paper with their undergraduate advisors and attended conferences to present a poster of the findings, neither of which I experienced.

From all those tree trunks, I learned the rudiments of field data collection, the anxieties of sampling design, and the process of formulating a hypothesis and then testing it. In the end, I looked at temperate forests quite differently—knowing their trunks are dynamic hubs of expansion and contraction, and the seasonality of leaf growth is fine-tuned with the commencement of trunk growth. A few decades later, field biologists like me would figure out that global climate change is making wood approximately 10 percent weaker because plants grow faster (and have less dense cell structure) with warmer temperatures; such changes in trunks invariably impact tree height, health, and growth. Trees are the biggest, oldest, and most iconic plants of all. Their planetary biomass exceeds 400 gigatons (GTs) of carbon, compared to only 2 GTs for wild mammals and a mere 0.06 GTs for humans. Their arboreal machinery is complex and interconnected from bottom to top, and for me, this initial foray into wood growth research ultimately piqued a curiosity about the whole forest, not just the trunks.

After college, I was determined to immediately enroll in a PhD program. My underlying reason was a secret fear of otherwise getting married and settling down like most of my high school friends. I had a wonderful college boyfriend, also a biology major, but he was heading to medical school. Why not move in together and get married? he proposed. I could eventually select a graduate program near his. I was torn. It would be a comfortable life, to have a husband with a medical practice and no need to ever work myself. But the trees had a strong pull on me; I was determined to pursue new botanical discoveries. And to accomplish that, I needed some experience outside the comfort zone of New England temperate forests. I cried. He cried. We sorrowfully parted ways, making the superficial promise we would get back together in a few years, although every couple knows this rarely happens.

The year was 1976, when tree-hugging environmental majors fervently believed forests would grow forever, outliving humans and providing stability to the planet if we only offered them protection via fencing or some technical government policy. The phrase "climate change" was not yet in our vocabulary and it never dawned on us that anything except clear-cutting, burning for agriculture, or the odd insect attack would threaten trees in a global fashion. Only two decades later, forests would face the ultimate threat of an increasingly inhospitable planet due to extreme heat, drought, wildfires, and insect outbreaks exacerbated by a warming climate. To nourish my tree-love, I applied to two forestry graduate schools, one of which offered me a full scholarship. My selection was made entirely through a financial lens since I was still juggling an undergraduate college loan. It seemed a miracle to receive a Duke fellowship. From a small-town girl's global view, I also thought moving to the piedmont landscapes of North Carolina was akin to seeing another continent, offering a new set of tall green subjects to study. Unfortunately, Duke University's forestry school in 1976 was not much different from my undergraduate geology experience. Only two women were accepted in a class of over thirty students. I eagerly signed up for some botany courses because the ratio in that department was more favorable, approximately 1:3. One my favorite classes was arctic ecology with an iconic botanist named Dwight Billings. Despite his enthusiasm for plants, he intimidated me, as did other classmates from places like Washington, DC, Santa Barbara, and Chicago. Could I hold my own with all these students from large universities who had coauthored publications with their advisors and navigated scientific conferences like pros as part of their undergraduate experience? They sprinted to the library every Friday night to be the first to read the newly arrived journals. Even worse, Dr. Billings required us to present two in-class lectures. I threw up in the ladies' room before giving both oral presentations that semester, overcome with shyness and totally tongue-tied. Meanwhile, in forestry classes we analyzed board feet of timber in loblolly pine plantations. To make our calculations, I learned a computer language called Fortran on an enormous IBM computer whose metal footprint occupied an entire classroom. Each student created a deck of punch cards to calculate different assignments. We carted enormous boxes of cards to and from class. Feeling out of my depth, and not really motivated by the concept of calculating board feet, I sought solace outdoors—in the Duke Gardens, or simply looking for piedmont wildflowers along the roadsides.

Running was still my favorite way to clear the cobwebs from my brain. I started running during college because it was a great excuse for tree-spotting. There were no girls' sports teams in high school so I did not come from an athletic background, but I guess running became another solo occupation, kind of like pressing wildflowers or measuring trunk growth. Duke had a jogging course that meandered through the woods close to the science building. I was a regular, and preferred morning runs when the birds were singing and fewer human footsteps were pounding the dirt. On a brilliantly sunny Saturday, I felt exuberant in the crisp air, running alone in the woods. It was a surprise to see a tall, athletic runner coming toward me at such an early hour, and we passed on the narrow trail. I thought little of it until I suddenly heard heavy breathing behind me. The runner had turned around and was gaining on me. Some inner sensibility came into play. I started looking for escape routes, just as his enormous hands grabbed my breasts. It would do no good to scream—no one was around—but I was a nature nerd and knew my way through the underbrush. With a burst of adrenaline, I dashed left into a tunnel of sumac, dogwood, and grapevine. I was not fast, but I was small, and dense vegetation was the best chance for escape. The assailant was gangly and over six feet tall, so the vines tripped him and the understory branches slapped his face and gripped his athletic limbs—Mother Nature's snare. I surprised the attacker with my zigzag escape and managed to outrun him through the dense foliage, sprinting all the way to my office. With heart pounding, I just sat at the desk and shivered for a full three hours before gaining the courage to dial campus security. They yelled at me, asking why I took so long to call. They explained a rapist had been reported on campus over the past three months. I asked them why any signage had not been posted. There was silence on the line.

It took this frightening experience to galvanize me into the radical decision that a male-dominated forestry school was not for me. I had amassed enough savings from working part-time at the Environmental Protection Agency in Research Triangle Park to sponsor my own sabbatical. Sometimes choices are not made strategically or with sophisticated long-term planning; this decision, triggered by a sexual assault, ultimately led me to seek a change all the way across the Atlantic Ocean. The University of Aberdeen, Scotland, offered a twelve-month master's in ecology, a program that had caught my eye in the Williams College career office more than a year before when I was looking for graduate school options. At the time, the tuition of $5,000 was beyond my financial means, and international students were not eligible for Aberdeen's financial aid; a year later, I had saved just enough from my part-time job at the EPA. An unexpected departure from Duke prompted no questions or exit interviews about gender challenges, assailants, role models, or lack of mentoring, which are common conversations in today's campus landscape. My resignation at the EPA, however, prompted them to tempt me with a full-time job. I had worked in the air pollution regulation division, a lone female among several hundred male engineers. The job involved reading big stapled regulatory reports, to compare the details of allowable air pollution between state boundaries and predict what conflicts might arise when, for example, one state had lax regulations but an adjacent state was more strict. In those days, such comparisons were made by hand, not using a sophisticated computer model as is the case now. Their offer put me at a crossroads—should I settle into a padded chair, gazing at air pollution regulations for an entire career? I could imagine a solid government retirement package at the end of four decades, but I was already bored stiff in a comfy office for hours on end. I couldn't envision a career path that didn't involve fresh air and forests. I relinquished the paycheck and enrolled in the master's in ecology program at the University of Aberdeen.

I packed warm clothes and hugged my poor parents, who felt like I was heading to the moon. Flying from Elmira to Scotland was a big emotional transition as well as a physical one. From the air, upstate New York was a complex mosaic of farm and forest, the only truly checkered parts of my past: those squares of farmland that turned white in the winter, interspersed with the dark squares of wooded terrain. This black-and-white pattern was especially visible because snow settled in the fields, not among the trees where the heat from living canopies melted the white stuff. In a figurative sense, the winter cross-hatching coloration of rural farms and forests was illustrative of the dichotomy between the tanking of my region's economy and the un-spoiled natural beauty that remained. I had seen the aftermath of a big flood during my senior year in high school, and five years later almost every factory or corporate headquarters in our region had migrated to locations farther south. Increasingly, you had to look beyond a creep-ing malaise of car bodies piled up in backyards to appreciate rural New York State's backdrop of gorgeous woodlands.

When my plane landed in Aberdeen, the sky was cold and gray, which I soon learned was the norm. A year in Scotland can be sum-marized by 364 gray-sky days, confirmed by my one thousand–plus photographs of sunless landscapes. Despite all the gray vistas, I had a truly life-changing experience as a graduate student overseas. The University of Aberdeen was situated along the shores of the North Sea, within sight of petroleum rigs offshore, which led to oil blobs peppering the beach. The presence of American oil companies in Ab-erdeen was both a blessing and a curse. Upon hearing my accent, most shopkeepers assumed I was a wealthy oil wife, but they quickly re-vised their thinking after seeing my dusty khakis, oilskin coat, and the dented thermos hanging out of my tired rucksack. To afford tuition on a modest budget, I lived low on the food chain and needed to find very inexpensive accommodation. Two classmates and I discovered an old farmhouse about fifteen miles north of town, available for free if we helped the owners with their barley harvest.

I had an upstairs bedroom where a jackdaw nested in the chimney and cold winds whipped down the shaft during big blows off the North

Sea. I could see icy gray swells from the window, just beyond the barley fields, which danced wildly in those never-ending northerlies. My best purchase in Scotland was an electric blanket, which cost a whopping five pounds. I wrote my entire thesis in that electrified cocoon, which literally saved my life in a cold farmhouse with no heat or hot water. I budgeted approximately five pounds per week for living, which barely paid for fish, cabbage, tea, and a few packs of biscuits (the Scottish term for cookies). My housemates were also on a stringent budget, so we teamed up to share groceries. One large Scottish cabbage served all of us for about a week, and I became an expert at using just the correct amount of salt in a boiling pot to bring out the flavor. Almost weekly, I ventured to the fishing docks to buy fresh catch, which was very inexpensive when purchased directly from the boats. My two housemates, Alan and Peggy, owned a dilapidated Morris station wagon, so old that moss grew in its metal siding, and they were savvy about finding roadkill for a main course. If a dead rabbit was cold, they surmised it died a slow death from myxomatosis, a disease ravaging the hare populations in the Highlands. If the rabbit was warm, they figured it had been in good health but hit by a car, meaning safe for human consumption. Alan wielded his sharp machete and behaved like a mad symphony conductor as he flailed it around our kitchen, butchering the carcass for stew. We added a few onions, boiled the meat for hours, and concocted a hearty meal. My children are embarrassed their mom ate roadkill as a budgetary strategy during her student days, and admittedly it was probably a bit risky. I would never advise them to do such a thing.

After classes commenced, I saw a tiny posting on the departmental bulletin board requesting a weekend guard for an endangered little tern colony at a seaside rookery (aka breeding colony) about five miles up the coast. The assignment was to prevent walkers and dogs from interfering with tern nesting. I was hired, bought an old bicycle, and spent each Saturday holed up in a sand dune from dawn until dusk with a thermos of coffee. The days were long and cold, with that North Sea wind whipping off the frigid water, but I loved watching the terns and hearing their raucous calls as they protected their eggs. The pay was five pounds for every Saturday during nesting season, which covered modest groceries and still left enough for me to purchase an occasional bus ticket for a hiking trip to the Highlands. Scotland boasted a weather pattern that could drive nearly anyone to drink. And so it did; I witnessed the locals in our nearby fishing village migrate to the pub around 4:00 p.m. as I shivered while riding home from the tern rookery.

Aberdeen University required a research thesis to achieve the master of science degree in ecology. I was immediately smitten by Scottish birches, reminders of an upstate New York childhood, which grew in the Highlands and were the toughest trees I'd ever seen. Gale-force winds and the prospect of snow almost any month of the year is daunting even to the most robust buds and foliage. As a result of this extreme environment, trees at the tops of hills were much shorter and scrawnier than in the valleys. I was curious about the phenology of birch along their elevational gradient, ranging from warm and almost subtropical in the sheltered valleys to freezing cold with arctic winds at the top. So I decided to survey seasonal variation of their leafing and flowering patterns at different elevations, which required an active schedule of hill-walking (as the Scots called it). My advisor, an enthusiastic tropical botanist, wisely spent much of his winter months in Malaysia conducting research on tropical trees. (He waxed lyrical about the mysteries of rain forests, so much so that his stories ultimately inspired me to find my own way to that part of the world.)

Over the course of a year, I befriended a crusty old forester who seemed to know the location of every single birch tree in the country. I met Richard when my Aberdeen class visited the western Highlands and toured some of the forestry plantations he managed for the shire. After that, he hosted me for some of my birch-spotting expeditions, and probably saved my life on several occasions as we tried to find patches of trees at the highest elevation. Almost every weekend, I took the public bus over to Skye or Inverness or Loch Ness, buying fish and chips in its traditional newspaper wrapping at a petrol station en route, and then met up with Richard to observe birch foliage in remote parts of the Highlands. I traveled light—a sleeping bag, a mountaineering tent, and a tiny stove that could boil a cup of tea in five minutes. Richard had his own gear, but best of all, he had an internal compass when it came to navigating in the hills. On occasion, we encountered a whiteout, which amounted to a dangerous blizzard when fierce winds whipped up the snow and enveloped the entire landscape in a white cocoon. Many hikers have died under these conditions, but Richard managed to guide me down from my high-elevation birches, walking blindly and numb with cold, but always reaching a valley where we could pitch our tents. Over time, I probably hiked several thousand miles in Scottish hills and examined hundreds of birch trees, taking careful notes about their leafing and flowering status, but mostly gaining a huge appreciation for their tenacity. It always felt great to remove my rough wool socks from my leather hiking boots after a long day of frigid hill-walking and squeeze my toes to make sure frostbite had not set in.

In addition to monitoring the phenology of birch canopies over the western Highlands, I was also curious whether individual trees leafed synchronously between bottom and top; perhaps the microclimate was warmer in the sheltered understory as compared to the windswept canopy? In the valleys, tree height reached thirty feet, but near the hilltops, only ten or fifteen feet. My advisor, Peter Ashton, thought this question was a great detective opportunity to study the whole tree instead of just the base like conventional foresters. He helped me collect a few old poles and planks, and together we crafted a rough, albeit hazardous, scaffold to survey the tree crowns about twenty-five feet high. I did not appreciate it at the time, but our makeshift framework launched my inaugural ascent into the canopy and lifelong career as an arbornaut. My first canopy research started with stunted Scottish birches and a rickety scaffold! I dragged the contraption to different hillsides using Peter's family station wagon, looking like a dumpster diver who had rescued some old timber and metal struts, but in fact these discarded bits of construction allowed me to measure buds in tree crowns. The scaffold methodology only lasted two months, after which time some university maintenance people, upon finding the pieces carefully stowed in a corner of the botany parking lot, dutifully dragged the metal contraption off to the university dump. But that was enough to provide me with up-close insights into a birch tree's seasonality from top to bottom, confirming that lower branches greened up before the upper canopy, thereby utilizing sunlight before the overhanging branches shaded them.

Field observations also showed that birch leafed out at least one month later at the higher elevations as compared to the valleys. Not surprisingly, trees on the hilltops were so dwarfed by the rigors of climate that they did not grow taller than ten feet, so their buds burst simultaneously throughout the crown. But in the sheltered valleys, where trees grew to at least twenty-five feet, budburst in the understory occurred two to three weeks earlier in spring as compared to the canopy. Although I did not appreciate it then, at the end of summer, temperate forests take a cue from seasonal day length to prepare for their winter "hardening." When the days grow shorter, trees strategically prepare for the cold by winterizing their entire machinery, so woody cells are not left with excessive water that might otherwise expand to burst the cell walls with those first freezing nights. Recently, as the onset of climate change creates increasingly great oscillations in extreme weather, the environmental cue of seasonal temperatures is proving less reliable, wreaking havoc on Mother Nature's systems. But the sun, whose light through the millennia has provided regular cycles of day length, also called photoperiod, remains a constant, signaling trees to shut down their operations for the winter and then ramp them up in the spring. If plants relied entirely on temperature, not sunlight, as indicators for seasonal changes, they would most certainly have suffered extreme confusion and more widespread mortality, especially during recent times with the rapid onset of warming trends.

From that short Scottish summer, I stumbled onto a novel observation that inspired future research. Aphids attacked the birches en masse just after all the new leaves appeared. The carnage was shocking— leaves were sucked, disfigured, and scrunched into dry, shriveled corpses. This was my first confrontation with six-legged foliage enemies, technically referred to as herbivores. As I observed with horror, aphids did not actually chew foliage but instead sucked their juices, leaving a rattling skeleton of dried leaf carcasses. Aphids infested 85 percent of foliage in the valleys, but only 35 percent in the exposed, windswept hills. So if you are a birch, maybe a good defense against insect pests is to live at the top of the mountains, where weather is too extreme for your enemies? I could have spent a lifetime in Scotland, studying the fate of *Betula* in valleys versus hilltops to understand the interactions of leafing, weather, and aphids. As with many aspects of ecology, one year was not enough to draw firm conclusions. And as I would later learn from tropical trees, even several decades were not always enough to accurately answer ecological questions.

Despite an entire year of miserable weather, I loved field research on Scottish birch. This was my first academic experience where gender was not a disadvantage, because Aberdeen's ecology course and faculty were extremely inclusive. We not only had equal gender representation in our class of twelve, but also students from five cultures. After hearing my Scottish advisor tell stories about his tree research in Malaysia, I became infected by his tropical bug. Peter Ashton was a world expert on a major family of trees called Dipterocarpaceae. These tall denizens dominated many forests in Southeast Asia, and he had written much of the definitive biology of these important trees. He told amazing tales of forays into Malaysian jungles, surrounded by flora and fauna unknown to me—ranging from sun bears to leaf langurs, hornbills, and slow lorises—and unexpected encounters with several species of deadly cobras. Hearing Peter's stories of sultry heat and dehydration in Malaysia while sitting in a Scottish pub with the cold North Sea winds rattling the windows was more than enough to inspire my daydreams about working in a tropical habitat. By amazing coincidence, several weeks later I met an Australian botanist spending his sabbatical at Cambridge University. Along with Belize, Australia was one of only two English-speaking countries with tropical forests. I learned from this chance encounter that the University of Sydney offered generous scholarships to international students.

Suddenly, my sights were set on the Australian tropical jungle, even though I was woefully unprepared. I applied to the program, was accepted, and joyfully purchased the cheapest ticket available, flying from

London to Sydney for $200 via People Express airline. I mailed the Aberdeen diploma back home. Mom cried when I called her from a pay phone. The notion of her only daughter moving even farther away from Elmira, New York, in pursuit of leaves was more than she could bear. We had a close mother-daughter relationship but somehow never shared life-changing decisions such as college choices, boyfriends, or what I wrote on college essays. My parents offered unconditional love and trust but left the choices completely up to me. I can't imagine how my mom would have reacted if she knew it would be thirteen long years before I lived on American soil again. I boarded the plane in London with forty pounds of books in a carry-on bag, trying to avoid paying for excess luggage. Dressed in wrinkled khakis and hiking boots, I was greeted with frowns from the stewardess who noticed my heavy science library. Most females boarding the plane had makeup kits or extra jewelry as carry-on, but I proudly carted botany books as I headed halfway around the world to study enormously tall plants.

In those days, a flight from London to Sydney took almost twenty hours and required a stopover for fuel. I sat awake, anxiously wondering what lay ahead. A gang of Australian blokes became drunk in the back of the plane and thought it was quite humorous to ask girls to take off their shirts as a toll to access the bathrooms. I was not amused and, in fact, started to realize I was about to immerse in a new culture. "Stone the bloody crows," those drunken Australians taught me to say when something crazy happens, which simply translates to a more vivid version of "Holy cow!" Our flight touched down to refuel in Kuala Lumpur, Malaysia, a place I knew of only as my Aberdeen advisor's winter tropical ecology retreat. This country's forests are among the tallest, dominated by dipterocarps, the most economically important family of trees in the world, yet I had never heard of them throughout undergraduate botany training.

Was I really cut out for graduate school in the "Lucky Country," as Australia was affectionately called? I did not even know what a tropical forest looked like firsthand, though I could still recall images in a *National Geographic* magazine I read during childhood. I did not relish the venomous snakes in a teeming mass under every canopy depicted in those glossy pages, and certainly was not an enthusiastic beer drinker like all the boisterous Aussies at the rear of the international flight. But even if I flunked out of the University of Sydney, it would be a small consolation to see a koala and add a new continent to my expanding bucket list of botanical wonders.

⫸ My Favorite Birches ⫷
(*Betula papyrifera*, *B. pendula*, and *B. pubescens*)

THE MAJESTIC WHITE OR PAPER BIRCH (*Betula papyrifera*) grows in the backyards of upstate New York and graces many New England forests and roadsides. Its characteristic white bark peels easily and forms an important resource for canoe-making by the Onondaga, Cayuga, and Seneca tribes native to my upstate New York region. A famous American naturalist, Donald Culross Peattie, paid tribute to birchbark canoes with the following statement:

> To any American of an older generation (now, alas, even canoes are being made of aluminum) there was no more blissful experience than the moment when on his first visit to the North Woods he stepped into a Birch bark canoe weighing perhaps no more than fifty pounds, but strong enough to carry twenty times as much. At the first stroke

of the paddle it shot out over the lake water like a bird, so that one drew a breath of the purest ozone of happiness, for on all the waters of the world there floats no sweeter craft than this.

Who doesn't love the sight of graceful paper birches swaying in a breeze? But watch out—they are shallow-rooted and fast-growing, often the first to fall in a ferocious storm. Their roots penetrate about two feet at most below the surface, and birch can't tolerate living under the shade of others. Hence, they are classified as early successional, meaning they grow quickly in the early stages of forest development but die out when overtopped by taller trees such as beech and maple that comprise later successional species. It is possible to play detective and age a New England forest by determining which species occupy the canopy and classifying their status as early versus late successional.

Betula timber is harvested for veneer, plywood, furniture, and firewood. Native Americans used this species not only for canoes, but also for baskets, baby carriers, torches, moose- and bird-call whistles, and mats. Medicinally, birch was used to treat skin ailments and dysentery, and to promote milk production in nursing mothers. In the spring, birches were often tapped because their sap makes flavorful beer, syrup, wine, or vinegar. During childhood, we enjoyed many crackling family fires fueled by their papery-edged logs. Like most temperate trees, *Betula* has its share of insect pests. The bronze birch borer threatens the trunks, while outbreaks of leaf miners defoliate the canopies, and several fungal species cause canker diseases. Peeling birchbark has long been a temptation for children, but when stripped from living trunks, the beautiful white bark never grows back. Instead, ugly black rings take its place, so it is always best to remove shavings of white bark from a fallen log.

Birches produce male and female catkins, technically defined as spikes of flowers. The male catkins appear during summer, starting out as buds in the axils of the leaves, and during the following winter they emerge as erect spikes visible in the leafless canopies. By early spring, they lengthen and become droopy, eventually flowering with each male floret contained inside a four-lobed calyx. The female catkins are thicker, their florets borne without a surrounding calyx, and covered by overlapping scales tinted light yellow with an occasional tinge of red, eventually turning brown and woody. Birch fruits bear tiny nuts in cone-like heads about one and a half inches long, and release wind-borne seeds that disperse widely and germinate quickly in sunny, well-drained conditions as an early successional species.

Like many trees, *Betula* has relatives that exist in other regions dating back to evolutionary times when the continents were linked. Across the Atlantic Ocean where I completed a master's thesis, Scotland is home to some birch cousins of our American species, including the ballerina of the family, silver birch (*Betula pendula*), with her dangling branches swaying in the breeze like a choreographed dancer. Europeans often call this beauty "lady of the woods." Another species, and also a focus of my field research in Scotland, is the hardy Highland or hairy birch (*B. pubescens*), which endures the rigors of high elevations as part of its range. Despite a double whammy of extreme weather and thirsty aphids, these small, rugged trees exhibit extraordinary hardiness as the predominant canopy sentinel throughout most alpine regions of Europe, not just Scotland.

Whether in North America or Europe, a few good birch trees with some bird eggs in their branches can be inspirational to any aspiring naturalist, as was definitely the case in my own childhood.

3

ONE HUNDRED FEET IN THE AIR
⫸⫷

Finding a Way to Study Leaves in the Australian Rain Forests

FEELING A BIT LIKE A GROWN-UP VERSION of Tom Sawyer, I eyed the target and took careful aim. Ready, set, fire. My jerry-rigged slingshot propelled its fishing line and lead weight up and over a high, sturdy branch of a coachwood tree (*Ceratopetalum apetalum*), some seventy-five feet overhead. Gazing up in satisfaction from the humid floor of the Australian rain forest, I hardly felt the infantry of leeches swarming up both legs and the sweat bees invading both eyes, or even spared a thought for the venomous brown snakes lying underfoot.

Believe it or not, I made that shot on the first try. The tree rigging method actually worked! Almost ten thousand miles away from any friends or relatives, teaching myself to scale trees with a homemade harness and slingshot, I was pretty scared. With the fishline catapulted over the sturdy branch almost thirty yards overhead, I next slid the nylon cord along its trajectory, and then the heavier climbing rope. I tied off one end of the climbing rope around the trunk of an adjacent tree, knotting it at least three times, which was totally overkill. I grabbed the free end of the rope and was ready to launch. Double- and triple-checking my harness and foot stirrups, I soon memorized a protocol for checking all my gear, almost like an astronaut before liftoff. After the safety inspection, I clipped my two ascenders onto the rope, making sure that the foot jumar was above the chest jumar, otherwise I would turn upside down. Squatting, I sat back in my harness and then slid the jumars up the rope, acting out the antics of an inchworm. Slowly, the ground receded and dense leafy foliage surrounded me. The dark-green leaves of the understory swallowed me up. Two of my caving friends, Al and Julia, watched from the ground, hoping I would remember all the pointers for safe use of their borrowed equipment. I hardly dared to look sideways let alone down, spinning in midair on a half-inch-thick lifeline, feeling akin to a tiny caterpillar ballooning on a silk strand through a huge expanse of green. I swayed back and forth, a bumbling first-timer with little sense of balance, flailed at the tree trunk, and grasped the rope for dear life. But as I climbed higher, moving upward became easier . . . practice, practice, practice. Beams of light began to flicker on my face as I drew closer to the top of the coachwood. Then mayhem broke loose around me. I had entered the sun-flecked leaves of the official upper canopy and encountered a sensory overload: creatures munching, flying, crawling, pollinating, hatching, burrowing, sunning, digesting, singing, mating, and stalking. The life surrounding me was nearly entirely invisible from the forest floor.

The tree extended another fifty feet—I was only ninety feet up!—into the teeming, buzzing hot spot of biodiversity called the upper canopy. I stayed in my aerial perch for at least an hour, which seemed like an eternity, awestruck by all the activities around me. I had ventured into a new world. I could hear the lyrical melodies of crimson rosellas, punctuated by the crack of eastern whipbirds in another treetop, but closer at hand were swarms of buzzing pollinators, colorful beetles crunching with their mouthparts on new leaves, and butterflies flitting in sunspots as they sought flowering-vine nectar for breakfast. I wasn't an entomologist—and even if I had been, I'd have had no idea what most of the creatures were doing there, since no one had ever been up here before! It was humbling to enter their world and think of how unknown all these creatures were to all of science, and more humbling still to realize my presence did not frighten any of them to fly away. I watched. I held my breath in wonder. I twirled on the rope to see in all directions. How would I ever make any sense of all this? I fumbled for my bulky camera, struggling to remove it from my backpack without dropping it or unclipping any safety gear, and snapped some photos, which I later realized were pathetic attempts to capture this new world. I was tempted to pull out a notebook and scribble some observations but couldn't do justice to everything around me; all I could do was gaze in awe. Eventually, I descended back down to the dark, relatively quiet and empty understory, giddy with amazement and feeling almost intoxicated.

I had a healthy respect for heights, although not an actual fear, so my climbing was cautious but determined. I was in good physical shape, but thanks to the hardware, I didn't need to be a superathlete to slither up the rope. In fact, I was not even out of breath after that first climb, because the ascending gadgets, the jumars, had angled teeth that slid up the rope (but not down) and held me safely in place whenever I stopped to rest. I soon learned to pause and enjoy the view, grateful that the equipment buoyed me safely in place. Still, the next day I was nearly bedridden, every leg and arm muscle aching, because I had instinctively tried to hug the tree trunk with both knees and frantically grasp all the branches with my arms, like a monkey. After a few ascents, I forced myself to remember this was not necessary, because the hardware and the harness provided support. Despite my exhaustion, the pure joy of seeing so much life among the canopy foliage electrified every brain cell, and my enthusiasm far exceeded any levelheaded realization of the scientific importance of those first climbs. I was, in a sense, reliving the thrill of discovery from a temperate forest childhood with its tree forts and bird nests, yet the tropical treetops housed at least tenfold more species to observe and appreciate. And instead of birch trees growing 50 feet high, these green giants rose almost 150 feet, which explains why their aerial commotion was completely inaudible as well as invisible from the forest floor. Finding a whole new world in tall tropical trees, I had taken my next step in evolving from a small-town nature nut into one of the world's first global arbornauts.

☙

Many months before I ever envisioned making a slingshot or borrowing ropes to climb tall trees, I daydreamed about my naive vision of rain forests: Tall. Green. Dense. Dangerous. Snakes everywhere. Jaguars lurking. Filtered light. Pungent decay. Butterflies flitting. Bird choruses. Back in the 1970s, few aerial images or whole-country surveys even recognized that deforestation was an insidious threat to these precious ecosystems. As a neophyte, I truly had no realistic sense of the height or complexity of this forest system, and Australia was about to swallow me up in her jungles for several decades. Gazing out the airplane window upon the Sydney Harbor Bridge at dawn, I not only envisioned tropical trees but also thought it was one of the most beautiful cityscapes I had ever seen. Clutching my passport, I schlepped my oversized bag of botany books off the plane and gathered up two small suitcases full of Scottish woolens that I hadn't known would prove useless in the Australian tropics. I was half a world away from any friends or family and embarrassed that I didn't even know what a rain forest looked like. Could I possibly achieve any success in demystifying forest secrets as a lowly graduate student in Australia? I was met by a fellow American student who ushered me to some temporary housing before he took off to the coral reefs for his own research. I don't remember much about those first days except that a possum peed from the rafters onto the bed where I was staying. I guess the Australian wildlife were welcoming me!

I slept off all my jet lag, and stepped out into the busy city, experiencing a few near-death street crossings until I remembered they drive on the opposite side. To a girl from tiny Elmira, New York, Sydney was like another planet—full of exotic trees (mostly gums), amazing bird songs and traffic noises, great public transportation, lots of parks and beaches, and a huge number of happy-go-lucky, fair-dinkum

Australians. After the cold winds and spartan living in Scotland, Sydney was a lush tropical oasis. I arrived at the university the following Monday—November 3, 1978, to be exact—and I went directly to meet my new boss, the head of the school of biological sciences. His office was in the Botany Department, housed in a classic, old-fashioned building of overly worn tiled floors and hallways that reeked of musty chemicals and were lined with old cabinets bursting with many decades of archives stored by professors long dead and gone. At that first meeting, I was immediately introduced to Australia's academic view of women. "Why is a nice girl like you wasting her time to do a PhD when you will only get married and have kids?" This was my first conversation with the head of the sciences, who was old enough to be my grandfather, and I was too stunned to reply. I was also secretly terrified he might be correct, but bristled at his narrow idea of a woman's role, all too reminiscent of my time in geology and in forestry school. Within the first day, I also met all the women in biological sciences; easy enough—they could be counted on one hand. There was one lone female assistant professor, two female graduate students, and a bevy of women secretaries and technicians working for approximately two dozen male professors plus their (male) graduate students.

I had come to the University of Sydney because rain forests grew nearby, and because of their generous international scholarships. This translated to three full years of funding toward housing and expenses, along with free tuition. No strings attached. Graduate students were not even required to teach or assist in laboratories. But beneath an outward excitement about encountering a whole new climate zone, continent, bird list, vegetation, and everything else that comes with it, I had gnawing doubts about whether I could achieve this lofty goal of a doctorate degree. I had applied to the Botany Department, proposing to conduct research in tropical and subtropical rain forests, ecosystems I had never seen, halfway around the world from any tree I had ever known. The year was 1978, and although inconceivable, tropical deforestation was not a critical issue as it would become forty years hence. It turns out that rain forest logging was just starting to ramp up in Africa, the Amazon, and Asia during the early 1980s, but aerial surveillance did not exist to monitor it with advanced technologies. And Australia, which had insidiously cut many of her trees already, leaving relatively tiny pockets of extant rain forest, was not even considered by most international rain forest surveillance. That was about to change during the course of my thesis.

Each day, the department hosted morning tea, which allowed the graduate students to mingle with one another, interact with faculty informally, and share information. When I was introduced as a new student who'd just arrived to study Australian rain forests, I sensed an invisible stone wall. No one in this tearoom of almost exclusively male faculty and students expressed a heartfelt welcome. I was on edge and disappointed, but there were a few other American students who later came to my rescue, explaining it was not culturally acceptable to warm to a "blue stocking" (Australian slang for a woman with intellectual pursuits). It soon became obvious I needed not only to achieve scholarly distinction as a tree scientist, but also to prove women deserved a place in this field.

After growing up among trees that lost their leaves every autumn, and then working on deciduous birches in Scotland, it seemed wise to focus on something similar yet different—in this case, tropical leaves. It may sound simple, but I was totally awestruck by the fact that most tropical trees stayed green all year long. As a result, their canopies had no clear sense of a defined seasonality of leaf fall. Did a new bud burst every time a leaf fell? Did each tree exhibit leaf expansion every month of the year? Were there some subtle seasonal pulses, impossible to detect at ground level? I was puzzled by this permanent greenness, and wanted to play leaf detective, comparing tropical canopies to my comfort zone of temperate deciduous foliage. Shifting ecosystems is challenging for any young scientist, but moving halfway around the world to Australia was a whole other kind of terrifying. Even at the best of times, I was pretty tongue-tied, immersed in a new culture and unknown landscape. At least Scotland had familiar species of trees from my childhood. The only thing I knew for sure was that I needed to keep my wits about me because so many creatures in the tropics were venomous.

My plan was to build on my Scottish birch research and address parallel questions about tropical trees. How long did leaves live in a tall evergreen rain forest canopy? What enemies threatened such long-lived foliage? Did these leaves suffer from aphid attacks like Scottish birch leaves? Within a month after my November arrival, I tried to make premature choices about which tree species to study and select field sites in different rain forest patches separated by vast distances. It was decidedly overambitious. My thesis advisor, a gentlemanly botanist who emigrated from England to study fire ecology, persuaded me to slow down. (He was not a rain forest expert by any means, but knew a lot about dry forest trees, especially eucalypts, which require fire to regenerate.) I had read a few publications about Australian rain forests, learning there were a whopping twenty-four technical types, all classified according to soil, vegetation, and geography by one of Australia's two experts. It was impossible to study all twenty-four in three years, but fortunately there were four general types defined by elevation and latitude: tropical, subtropical, warm temperate, and cool temperate. The selection of field sites is critical for research. My advisor offered to provide a vehicle and driver if I would visit a few sites before finalizing any decisions. He assigned a reluctant botany student to help me navigate the logging roads on a pilot expedition to visit those four rain forest types in northern New South Wales and southern Queensland.

I was sleepless in anticipation of getting into the bush, the Australian term for any dense vegetation scrub. Visitors are often surprised to learn Australia is only slightly smaller in land area than the United States, with 2,969,907 as compared to 3,794,100 square miles. Rain forests only exist there along a narrow coastal band of up to fifty miles inland, where prevailing winds from the Pacific Ocean bring the necessary rainfall to Australia's eastern slopes. The escarpment blocks rain from moving into the continent's interior, otherwise known as the outback, those many hundreds of thousands of square miles where a few cattle and many kangaroos eke out a life on arid pastures. Australia's vast interior is dotted with patches of dry forest, composed mainly of the genus *Eucalyptus* (also known as gum trees). Rain forests occupy only about eight million acres of this vast country, a mere 3 percent of all forested lands but arguably the wettest places; they house 60 percent of the country's plant species, along with 35 percent of mammals and 60 percent of bird species. At that time in the late 1970s, most small forest stands became slated for logging. The national determination to clear rain forests ultimately impacted my student years as a decade marked by environmental controversy, poaching, and political demonstrations. But to make a long story short, 32 percent of those remaining stands are now conserved as UNESCO World Heritage sites, a source of pride for most Australians. Many former loggers reluctantly transitioned into ecotourism operators but are now millionaires with sustainable incomes.

To make my field research more manageable, I needed to select a subset of important tree species within each forest type for long-term leaf investigation. The decisions about what species, which sites, and how long to measure are the heart and soul of good fieldwork. For any research that relies on repeated observations over several years, it is tough to start all over midstream, so it's always best to design date collection carefully from the outset. And the rigor of research depends upon designing a good sampling scheme. I decided to compare leaf dynamics of three species of trees at three sites in each of the four major rain forest regions. The regions were: (1) subtropical—warm, wet, and resembles true tropical rain forests but with fewer species due to its location slightly farther away from the equator; (2) warm temperate—warm and wet but located in temperate latitudes, resulting in lower species diversity than the subtropics; (3) cool temperate or montane—moist and located at the tops of ridges, housing lower species diversity; and (4) tropical—the most diverse type, located closest to the equator and at least a two- or three-day drive from Sydney.

To prepare for the field, I spent a week in the botany library reading everything about Australian rain forests. Unfortunately, only a handful of scientists had ever studied this ecosystem. One botanist had created the technical soil-based classification of twenty-four rain forest types, and another made a guide to seedlings and their taxonomy. A third wrote several tree identification guides, and a fourth ecologist was based in California but traveled yearly to Australia to monitor rain forest and coral reef species diversity. It did not take long to read all their publications and realize my research was going to be relatively lonely. The existing literature confirmed that most tropical biologists were working in Panama and Costa Rica, which made sense—those countries were short flights from American universities and had major field stations with comfortable amenities like air-conditioning and dining rooms. Australia was either too far or too unknown to attract the same level of scientific curiosity from well-funded US or European universities. That worried me a great deal. Having no cadre of fellow students, no library of extensive research findings, and relatively little funding for rain forest research meant I was truly on my own. It was also increasingly apparent that I needed to account for the risks of conducting field research in forests that were rapidly being cut down.

Although I am fairly frugal with my travel wardrobes, I spent hours packing for this first rain forest expedition. Those well-worn Wellington boots from Scotland came in handy, as did several pairs of khaki pants, multiple long-sleeve shirts, rain ponchos, flashlights, and camping gear. No one advised me about what to take, but I had enough field experience from Scotland to figure it out, mostly. Even though Australia was stinking hot, I packed long-sleeve shirts to minimize bug bites. I bought several field notebooks and extra Kodak slide film. I still had my tiny kerosene cookstove, which almost fit into a jacket pocket and had literally saved my life by providing hot soup during many frigid outings in the Scottish Highlands. So I did the same thing for the Australian bush, packing a supply of dried soups, spaghetti noodles, oatmeal, and other easy camping meals. Australians take a "smoko" in both morning and afternoon, which originated as a cigarette break but in more modern times consists of tea and sweets. So I made a point to buy different Australian cookies to keep the driver energized. From a year in Scotland, I was adept at driving on the left side of the road, but in Australia it was still very much a man's place behind the wheel. We drove out of the urban congestion of Sydney, and then headed north all day before turning off on a logging road. My driver and fellow student claimed he had the map in his head, so I really had no way of keeping track of where we were heading, just "to the bush."

In late afternoon, somewhere between Sydney and Brisbane, the vegetation changed suddenly from the silvery blue-gray of gum trees to lush emerald green. We turned onto a very remote logging road with massive trees greater than four feet in diameter, soaring over a hundred feet high, and dense, deep-green foliage on either side. I gasped for joy, and had to pinch myself, because this was the real deal. The trees were not only tall, but shaped like lollipops, with a dense clump of foliage at the very top and festoons of vines and epiphytes draping down the trunks. This funny shape made sense, given the extreme competition for light in such highly dense and diverse forests. Tropical trees were in a constant race, and taller trees won out because their crowns received direct sunlight. I strained to make out the shape of a leaf, but they were too far above our heads. The joy was short-lived, however. My driver careened around the next curve and plunged the front of the jeep into a mucky stretch of slurry mud so deep that the front wheels sank out of sight. He cursed and tried to accelerate, but our tires spun deeper into the brown slick. On the first afternoon, I learned a lot of Aussie expletives, until my driver insisted I walk farther into the sub-tropical scrub so he could swear more freely in my absence. This was 1979, years before cell phones and GPS devices accompanied every expedition.

On first view of the rain forests, I was gobsmacked (and also frightened) by their height and inaccessibility. In Australia, the canopy extended from fifty to two hundred feet, impossible to calculate look-ing upward from ground level except when a road was cut through the jungle, exposing the height of trees in cross section. The abundance of greenery defied all expectations—if there is a place where oxygen is most abundant and pure, the rain forest's upper reaches would surely win, hands down. I looked up—green. I looked on either side of the trail—green. I looked down—brown, due to the decaying remains of (green) leaves on the forest floor. In short, this surround-foliage world fulfilled my wildest dreams about studying leaves, but several logistic constraints became immediately evident. First, how in the heck could

I reach the leaves when the canopy was so high overhead? And second, could I navigate through these tall, dense stands of trees that looked alike and seemed to have almost no trails? I also saw plenty of hillsides just recently logged and located adjacent to gorgeous primary (meaning original) stands that would probably be next. I was infuriated, but also inspired to get to work. I would soon learn that Australian rain forests were disappearing more quickly than their secrets were discovered.

My advisor was right to send me into the bush before I finalized any research plans; this first trip allowed me to set lofty yet realistic goals. And now I wanted to be up high with those millions of green machines (aka leaves) taking in sunlight. I returned to the botany library and eagerly read about how forest scientists had accessed tropical trees in other parts of the world (mostly cutting them down, I soon learned), and attended seminars to hear other students explain how they selected field sites (they all made it sound so easy). Immersing myself in the collections of the University of Sydney herbarium, I learned the fundamentals of identifying rain forest trees, although most specimens were restricted to just understory leaves in a two-dimensional state of brown, squished, and dry. Names flooded my brain after poring over hundreds of pressed plants, including genera I could hardly pronounce: *Acmena, Doryphora, Dendrocnide, Elaeocarpus, Sloanea, Orites*, and a major tongue twister, *Pseudoweinmannia*. I furiously wrote notes and started a series of notebooks on tree species, geographic locations, and field methods. I tried to give myself a crash course on Australian rain forest ecology, but next to nothing was published, and absolutely zero about the canopy or its leaves. Existing scientific publications in other tropical regions were similarly limited to the understory or, in a few cases, based on a fallen tree. Could I ever develop a worthy methodology to study rain forest leaves of the whole tree, which I now realized extended over a hundred feet high? How could I select which of hundreds of tree species to study, what geographic locations, for how long, and how many? There just was not enough existing literature or advice to help me narrow the focus.

I made friends with some graduate students in the marine biology department, who were a veritable gold mine of information and advice for designing fieldwork, although in their case for either intertidal populations of barnacles and their predators, or coral reefs with all their complex ecology. In particular, the coral reef students faced similar challenges of trying to narrow their field research amid a huge array of biodiversity. Like me, they were tempted to do too much. Unlike laboratory studies, where theoretical questions were often answered using one or two species in a cocoon of controlled conditions, field biology was fraught with obstacles: weather, floods, tree falls, bugs, drought, fire, edge effects, human activities, and sampling bias, as well as some factors that I forgot to consider. I probably should have received an honorary degree in barnacle population dynamics or butterfly fish ecology given the hours I devoted to assisting fellow students with their research. But our many discussions enabled me to design fieldwork more accurately and efficiently in the trees. Monitoring tropical fish in the three-dimensional habitat of a coral reef was not too much different from observing insects in a three-dimensional tree crown, except instead of tidal currents and sharks, I had the challenge of gravity and snakes. On top of everything else, as the only rain forest ecology graduate student and one of two females out of approximately twenty-five students, I needed to succeed in a male-dominated, marine-centric workplace. Fortunately, I met some wonderful male graduate friends and probably received more (not less) attention as the only student focused on rain forests. As a case in point, I met Hugh at the photocopy machine one Friday afternoon, where each of us was furiously making copies of the latest article by the world-famous ecologist Joe Connell, from the University of California, Santa Barbara, who studied species diversity in coral reefs and tropical forests. Hugh saw me clutching a copy of *The New Yorker* magazine and smiled. A *New Yorker* subscriber himself, he already knew that only one hundred subscriptions existed in Australia, which he commented made us both special. We spent the better part of the weekend discussing the Janzen-Connell hypothesis, a subject that inspired many of our fellow students to test hypotheses about biodiversity. Connell had established long-term plots in coral reef patches and on rain forest floors in Australia; over time, he monitored the success of each species, ascertaining important information that competition in higher diversity ecosystems actually sustains the system, rather than driving it toward single-species dominance. Both Hugh and I were extremely interested in diversity, particularly as it might impact the health of different intertidal ecosystems or rain forest trees, respectively. Hugh studied the seasonality and competition of barnacles along the New South Wales coastline. He transplanted different species of barnacles on artificial plates but carried out his work in the field with waves crashing all around him, almost unheard of, but successful in terms of results. Such chance interactions during those halcyon days of genuine intellectual inquiry were the stuff that made graduate school one of the best chapters of my life.

As I chattered with greatest animation to Hugh and other students over those first weeks, I began to crystallize research questions and field protocols for Australia's tall rain forests. So when that much-admired American ecologist actually showed up at Sydney University to give a seminar on his innovative research about species diversity, he asked if anyone was working on rain forests. Only one hand shot up. He needed a field assistant to identify his trees and seedlings, so I instantly had the job—no competition. I was walking on air at the prospect of working with this distinguished scientist, and ultimately our partnership lasted well over a decade. Like me, Joe Connell had worked in Scotland and now shifted his field research to Australia. For many years, he mentored me, and even gave me the ultimate nickname, Margaret Number 2 (his wife was Margaret Number 1), because we shared so many adventures and thousands of hours identifying seedlings and trees. (Someday, I may write another book on giving blood to leeches as we "groveled"—Joe's pet term for our field behavior—on the forest floor marking thousands of seedlings . . . I sometimes refer to such ground-based research as my retirement plan for when I can no longer climb.)

My prior experience with temperate leaves inspired some of my first questions about tropical foliage: What triggered these leaves to fall in the absence of cold winters? How did the ones at the bottom of the tree survive in such low light? And what about the uppermost foliage in that hot, relentless sun? These questions evolved into a thesis plan:

to observe leaves emerge, grow, and die. (I secretly hoped it would be simple.) Compared to the seasonality of wildflowers and trees in upstate New York, where everything shut down like clockwork after the first autumn frost, I needed to watch for other cues that triggered the end of a tropical leaf's life. All living things, evergreen leaves included, have a finite lifespan, defined by physical and biological factors. From reading, I knew most tropical trees were evergreen, meaning their leaves didn't fall all at once, but only some at a time, and so I hypothesized that leaf emergence similarly occurred throughout the year, without any well-defined season. It seemed pretty logical. Given the sauna conditions of a rain forest, I also figured a flimsy leaf couldn't last too long, flailing on the end of a delicate petiole in such a sultry environment with frequent monsoon blasts. So I took a wild guess that two years was as long as rain forest leaves lasted. The hypothesis I planned to test was simple: evergreen leaves emerge throughout the entire year, but each has an average lifespan of two years. Next, I needed to design field methods to challenge the assumption. I imagined myself lounging in a jungle hammock in a beautiful tropical setting, watching leaves fall over the course of several years, and then writing up a dissertation. But when I casually proposed this scenario to my advisor, he chuckled politely. He liked the hypothesis about leaf longevity but wasn't convinced I should passively record leaf fall as an observer on the forest floor. He suggested any worthwhile fieldwork required scaling the trees to examine the leaves where they grew. I was not athletically inclined so first proposed a few ground-based options that indirectly embraced the canopy region: I could train a monkey? Use super-powerful binoculars? Sit on a ridge alongside a gulley adjacent to some upper crowns? Use a shotgun to bring leaves to the ground? No, my advisor explained, if you want to study the leaves, then you need to access the whole tree, not just the understory. And if you want to study their lifespan, then you need them to stay attached to the tree. As an expert on dry vegetation, especially gum trees that often averaged around ten yards high, he had no arboreal skills to share and seemed blissfully unaware of the height of some of these rain forest giants. Although not a rain forest expert, he was the only botanist at the university who studied trees. True to his function, my advisor gently directed my enthusiasm into a solid hypothesis served by reliable methods to collect accurate data. He forced me to think creatively about fieldwork—starting with a clear hypothesis, and then working backward to establish how I could gather accurate data to test it. First, I wanted to study the leaves. Second, if I was going to monitor leaf growth in tall trees, I'd need to scale them—frequently. The problem was obvious. I just needed some strategies to reach those aerial subjects. By serendipity, I had already found the university caving club for advice and equipment.

My introduction to spelunking came during my first month at Sydney University. Funding was available for ecology graduate students to attend a conference in New Zealand. As the lone rain forest student, I was lumped into the scholarship pool with all the marine students. Mike, the American graduate student who had met my flight arriving into Sydney, proposed we could rent a car and see a bit of New Zealand after the meeting. At the conference, I was all ears to hear how different researchers set up their fieldwork, ranging from mapping kelp beds to counting coral reef fishes and trawling for tiny phytoplankton in the water column. I sat spellbound in the audience, daydreaming about my own experimental design and how to sample in an aerial three-dimensional space, instead of a watery one. After the meetings, Mike and I were supposed to meet up for departure in our rental car, but he was nowhere to be seen. I asked a few others, who chuckled and said, "Try room 122." It turns out he had met a girl at the final banquet, and they hit it off. When I politely knocked on the motel room door, he awkwardly threw me the car keys and said, "See New Zealand on your own." Suddenly, I was exploring a new country as a solo venture. I was fine driving on the left side of the road, having adjusted in Scotland, but a little apprehensive about camping alone in strange settings. However, not only did I hike alone in Tongariro National Park and camp near a hot spring, but I also stopped to see the notorious luminescent glowworms at the Waitomo Caves. By chance,

I ran into the park manager, and we traded credentials. He was about to lead a caving adventure to find ancient moa bones, a flightless bird that became extinct around the year 1300 due to hunting by the Maoris. He invited me to participate, saying more eyes were an asset to their nocturnal quest. I was not a speleologist, but it still sounded like fun. After donning a helmet and taking a quick one-hour lesson in harnesses and descending on ropes, off I went for an all-night caving expedition. I don't remember much about the actual descent because I was terrified in the dark. But we found lots of bones at the bottom of the cave, and this introduction to a vertical transect on ropes came in very handy a few months later as a neophyte arbornaut.

After spelunking in New Zealand, I sought out the cavers at Sydney University, who laughed at the notion of climbing up given they always climbed down. But they realized I had a serious determination to reach the treetops. Fortunately for me, the Waitomo Caves experience gave me confidence that caving equipment might be adaptable to the needs of an arbornaut. Arborists, those stalwart tree climbers who prune and cut urban trees, use a completely different outfit composed of heavy-duty gear suited for overpowering the branches and crashing through the tree crowns; their gear would not help me navigate a tree delicately, keeping every leaf intact. But cavers used lighter gear, conducive for carrying many miles on remote trails to access underground exploration and often worn all day on a long expedition. The Sydney University caving club made their own gear since no commercial recreational equipment was available in those days. One of the only female members, Julia, loaned me her industrial sewing machine plus a length of bright-orange webbing (probably obtained from a military supplier, since car seat belts were not yet available), so I measured my waist and thighs and copied their design to sew a basic harness. I never imagined I'd be grateful for my seventh-grade home economics class, where I became an expert operator of the original Singer zigzag sewing machine and had even mastered the art of making an entire pantsuit, including a zipper. By comparison, making a climbing harness was much simpler. But in addition to a harness, I still needed some technical hardware, so Julia's caving partner, Al, kindly sold me two jumars (toothed metal devices for upward ascent on a rope), a few carabiners (metal climbing clips) to connect ropes and hardware, and a whale's tail, which was a metal rack with four holes to thread the rope, thereby slowing my speed of descent. This was the perfect gadget for a leaf lover since I would never come crashing downward. I was also inwardly quite scared about dangling from a rope on a branch of unknown strength, so any hardware that slowed me down or provided extra control was much appreciated. Al also taught me to tie a few essential knots, including the climber's all-time favorite Prusik knot and the highly secure clove hitch. Fortunately, many of them stayed affixed to the gear so never needed to be retied. Last but most important, I needed a climbing rope that had to be more than twice the length of the highest climb, so the rope could be placed up, over, and back down from the highest secure branch. With just over two hundred feet of rope, I would be able to climb any tree approximately one hundred feet high, leaving enough extra length to tie off around a nearby tree trunk at ground level. Based on eye estimates from initial visits, I figured this was a good height for starters. I also bought a bicycle helmet, bright orange to match my carrot-colored harness.

But how would I rig the rope in the upper branches? Cavers simply drop lines down into a dark hole; their biggest hurdle is a lack of light for navigation. But I had to get my rope to literally fly, so I needed a slingshot, the only gadget that could propel a rope and also be used safely by a neophyte like myself. I soon discovered they were illegal to purchase in Australia, so I would have to make one. Most field biology students are constantly rigging gear, like metal frames for fish counts or special cages to capture small mammals, so I had already befriended the staff at the university workshop. A gray-haired veteran of gadgets named Basil helped me find a metal rod of the perfect diameter, and together we welded it into a classic Y-shaped catapult and cut a piece of old car tire rubber for the elastic. Then I tied a fish sinker to a reel of fishline and fired a first shot over a fifty-foot-high branch outside of the botany building. It worked. Out in the rain forest a week later, it was a different story. All the vines and dead branches in the tree crowns intercepted the fishline's trajectory on more occasions than not. Practice, practice, practice! Over time, I became the Botany Department's version of Tarzan's Jane (not good ol' Deadeye Dick) and learned to locate a solid branch with clear airspace beneath, which allowed for vertical ascent up the rope. In addition to the branch and the airspace, my vertical rope transects needed to pass through foliage at different heights that would be conducive for sampling. The process of finding the correct tree species, a safe canopy branch, and also access to leaves along the rope's pathway was a new skill set I needed to master in the forest, both quickly and safely.

After identifying a limb in the desired species of tree with the right amount of foliage distributed, I was ready to use my newfound slingshot. Rigging the climbing rope was a three-step process: first, propel a fishline via slingshot; second, attach the fishline to a nylon cord and haul it over the same pathway; and third, hoist a heavier climbing rope into place by tying it to the nylon cord, pulling it up and over the support branch. Once the climbing rope is in place, one end is used by the arbornaut and the other is tied off to an adjacent tree. Voilà! A vertical transect through the canopy is ready for access. Over the next few months, I refined my field gear through trial and error. I created a mold for melting lead, to make ideal weights shaped to propel a fishline through dense vines, and I adjusted the elastic slingshot bands to exactly the right length and width for shooting prowess. I also fabricated a special waist belt to hold gadgets—pencils, notepad, waterproof markers to label leaves, duct tape to mark branches, camera, and Oreo cookies for survival. And finally, I designed a hat with face mesh to block sweat bees and other swarming critters while I dangled for long periods of time. I did not tinker with the harness, even though it really dug into my butt, because I was too impatient to get into the trees. Looking back, I realize that I valued precise science a whole lot more than personal comfort. The harness, slingshot, metal climbing hardware, and ropes all fit into one duffel bag and soon became all I needed to access almost any treetop around the globe. The final, and perhaps most critical, ingredient for my toolkit was accessing a lot of inner courage I never knew I had, which only became known to me, literally, on the rope. And so that was how I found myself dangling in a coachwood tree as a neophyte arbornaut, cautiously climbing into the upper crown to discover a cacophony of biodiversity. Much to my surprise, many fellow graduate students (all male) wanted to accompany me into the field to rig trees, because the slingshot was such a fun gadget. Some shooters were successful on the first shot, whereas others took many expletives to get a line over the correct branch. There were no handbooks on climbing, at least none in Australia; there was only the caving club for advice. But if I could safely reach leaves throughout the whole tree, then my challenging research questions could be answered.

With the equipment list squared away and a beginner's sense of the logistics for scaling tall trees, I still needed to think harder about what trees to climb, how many, and why. Field trips into the bush to locate study trees required booking a university car, packing a full set of camping gear, scrutinizing maps to locate rain forests, and learning the nuances of driving in rural Australia—not just keeping to the left side of the road but also anticipating kangaroos, logging trucks, and mud. On the second trip, I happily went solo and visited three national parks—Dorrigo, New England, and Royal—all primary rain forests in New South Wales safeguarded from poachers and clear-cutting due to their preservation status. On the first day, I set up a meeting with a retired Dorrigo forester whose name peppered the field guides in the botany library. Alex Floyd had authored the only identification guides to Australian rain forest trees at the time. I sent him a handwritten letter, as was the conventional communication of the late 1970s, and he replied to confirm our date to look at trees. I was especially eager to identify species in the field, and Alex kindly agreed to give me some initial pointers. We met in the parking lot at Dorrigo National Park in the pouring rain. It was easy to find him because no one else was there. In Australia, the only remaining rain forests to have survived a century of logging tended to be stands on extremely steep hillsides where the timber was too difficult to extract. So we two soggy botanists proceeded to slog along slippery slopes and try our best to look up at silhouettes of different trees while constantly wiping raindrops off our glasses. Due to our shared passion for rain forests, Alex became a loyal colleague, perhaps because there were only a handful of us throughout the entire country who learned to identify these trees. (I later taught botany to his son when he was a college student!) Starting at the edge of the parking lot, Alex taught me to identify sassafras: he crushed a leaf so I could appreciate its beautiful scent, like that of the North American sassafras growing on the forest floor of my childhood; the Australian version was a tall tree, though, not a small understory plant. The aromatic scent turns out to be the only similarity between these two species, which grow half a world apart. (Here was a future research project, to investigate how sassafras aromatics evolved separately in a temperate understory plant and a tropical tree.) Next, we confronted the red cedar, highly valued because of its outstanding timber. Popular for furniture making, it earned local notoriety as the species responsible for the clear-cutting of Australia's rain forests (although people enacted the deforestation, not red cedars). And as one of Australia's only deciduous trees, it piqued my leaf-loving curiosity for research potential, although Alex cautioned me it was quite rare due to continued poaching.

As we walked through the forest, blithely staring upward to the treetops, then looking down to plant our boots on the steep, muddy trail to avoid slipping, I saw blood seeping through the fabric of my bedraggled shirt. Had I been cut somehow? I cautiously peered inside. Nestled between my breasts was a slimy black, puffy creature over an inch long. I had no idea what was attacking me. Terrified, I could hardly speak and reluctantly described this invader to my senior colleague. He chuckled and said, "Just a leech." Alex was very blasé, but I was horrified. In ignorance about the tropics, I had never heard of leeches living in the Australian rain forests, nor did I know how to remove one from my chest. Alex suggested to just leave them alone, and once they had a full blood meal they would drop off; that was not very reassuring, especially when a large leech was feeding on my chest. But by the end of the day, I had given a blood meal to at least a dozen leeches and eventually stopped concentrating on their invasion—Alex too had attracted his own population. Welcome to the Australian rain forest! Finding many more of those slippery suckers on diverse parts of my anatomy at the end of a wet day, I disgustedly flicked them out the window of the roadside motel. Even though their bite and subsequent bloodletting were not life-threatening, their attacks were a personal affront and hence I developed a lifelong mission to minimize their attacks. My initial combat method was to simply brush them off once discovered. But tossing them as you walk is considered impolite bush behavior, because they usually latch on to the person walking behind. Over the years, I resorted to more creative modes of leech removal: burning, cutting, salting, squishing, and even sewing canvas boots to my pants to prevent invasion, since leeches usually crawl up shoes to socks and then inside pant legs to the groin, seeking the warmest and coziest body crevices to suck a blood meal. As with so much rain forest research, I ended up with more questions than answers: Why are there no leeches in the American tropics? Does another organism occupy their ecological niche? Do these creatures inhabit the Australian canopy, or just the forest floor?

Alex introduced me to several dozen tree species, with handy tricks for field identification such as the smell of sassafras leaves or the red flowers of flame trees (*Brachychiton acerifolius*). Sometimes a little knowledge can be dangerous, and this field trip provoked many sleepless nights trying to sort out which trees would be both safe for rigging and offer unique foliage traits to study. As a solitary researcher and a woman, I also needed to think strategically about safety in the field. Australian rain forests are a pretty wild and remote region of the planet. Admittedly, I felt slightly apprehensive about driving, sleeping, and exploring alone, but simply had to buck up courage and move beyond those small-town fears of the unknown.

A discussion phase is critical for designing effective research. How could I mark branches overhead and find them again month after month? Could I camouflage each climbing site from human marauders? Which species offered the most intriguing leaf traits? Which trees were safest for climbing? Those friends in the caving club had been helpful with preliminary logistics, but now I required insights into the science. So I turned to the gang of coral reef students located next door in the Zoology Department. In terms of high diversity, rain forests are the land-based counterpart to coral reefs. The marine students and I found plenty to talk about because we shared similar questions about biodiversity and how to sample accurately in our complex ecosystems. Whereas I wanted to figure out how many leaves and at which heights I should sample to document foliage, they needed to determine how many fish on which patches of coral reef would answer questions about marine ecosystems. I needed an experimental design (essentially a recipe) for collecting field measurements over several years and then analyzing the results. But I also had to contend with different safety factors—snakes, leeches, muddy roads, biting ants, rotten branches, finding drinking water, worn ropes, and other unknowns as compared to the protected cocoon of an indoor laboratory. Like coral reefs, the canopy world was uncharted, so each element of research was more than simply a student exercise—it was a venture into the unknown. I probably spent half my time designing new methods and the other half doing the actual field research.

From my reading, I also learned that by the time I came on the scene in 1979, less than 10 percent of the country's original rain forests remained. Governments like the one in Queensland were hell-bent on logging them for timber extraction, without any understanding of issues such as biodiversity, extinction, or restoration. Of the estimated ten million square kilometers (6.2 million square miles) of the world's rain forest, less than one quarter of 1 percent existed in Australia (about 22,500 square kilometers, or 14,000 square miles), but it contained species found nowhere else on Earth. Less than 10 percent of that tiny percentage was safeguarded in national parks, so I felt a sense of urgency to explore those last remaining stands before they too disappeared.

I selected five tree species for long-term observations about leaf growth and death, a large-enough sample to ensure diverse patterns yet small enough for one arbornaut to manage. Each species was either representative of one of the major rain forest types and/or had an unusual foliage trait. They included:

1. Giant stinging tree (*Dendrocnide excelsa*), with extraordinary dense, sharp, stinging hairs on both leaf and petiole surfaces, which inspired my curiosity about their apparent defense armor to defy any hungry herbivore.

2. Red cedar (*Toona ciliata*), one of the few Australian rain forest trees exhibiting a deciduous leafing pattern, raising questions about whether its entirely different phenology had any survival advantages as compared to neighboring evergreen species.

3. Sassafras (*Doryphora sassafras*), the only tree species common to all Australian rain forest types, giving a chance to examine how its foliage adapted to different environments.

4. Coachwood (*Ceratopetalum apetalum*), with leaves that were elongated, smooth, and waxy, which was typical of most rain forest species, so it seemed important to study the "average leaf."

5. Antarctic beech (*Nothofagus moorei*), which defied the conventional rules of rain forest tree diversity by growing in mono-dominant stands, leading me to wonder why insects didn't overexploit such a homogeneous leaf supply.

I returned to my first coachwood for a pilot test of the field data collection methods and carefully enacted a protocol marking replicate leaves on three branches at three heights. I climbed to the first level of foliage, called the understory, which is deeply shaded and extends from zero to thirty feet high. I dropped a special, super-large forestry tape measure to record the exact vertical distance from the ground and repeated the same procedure for mid-canopy (ranging thirty to sixty feet high with intermittent sun flecks, small spots of sunlight streaming through the upper leaves), and then uppermost canopy, evident from the abundance of sunlight hitting the leaves (usually located above sixty feet). At each level, I selected three branches, each with five to fifteen leaves, and marked the first one by encircling it with an unobtrusive necklace of fishing line and yellow electrician tape labeled "Branch 1–1." In a small yellow waterproof notebook, I transcribed the meaning of each label: branch number 1 and height 1 (for example, five yards). Each leaf was numbered, and represented data points at each height, so I could monitor changes throughout a leaf's lifespan. I repeated this for branches numbered 2 and 3, also at the same height and light levels but simply replications.

The tricky part during the first climb of every tree was writing a legible number on all the leaves of each branch while dangling from a climbing rope. I had to avoid breaking a leaf when touching it or exerting undue pressure on a branch, even if wind swayed the rigging. The leaf at the base of each branch was numbered 1, and then 2 onward to the highest numbers toward the branch tip so I could continue the numbering sequence when new leaves emerged the following season. As with all plant growth, the new buds are at the tip of the branch, and so the new leaves emerge at the outermost region of the canopy. It required occasional acrobatics to reach along each branch from a sway-ing rope or sometimes flailing upside down in the harness to gently grasp the outermost leaves. I soon became adept at leaf marking and to this day have never broken a leaf or branch during many decades of treetop surveillance. But those first branches created great anxiety as I learned the nuances of how to make notes while dangling from a rope. One notable episode was my monthly visit to coachwood branch 2–2, situated in the mid-canopy of tree number 2, where I learned to push away from the tree trunk with my left foot, gently grab the branch with my left hand, and then lie almost horizontally in the harness to reach the newest leaves, which were growing farther away from the original rope transect.

During those first weeks of fieldwork, I bought a cheap orange mop bucket that became a mainstay, holding an ever-growing cadre of gadgets that had outgrown my waistbelt: waterproof Magic Markers, notebooks, pencils, rulers, graph paper, yellow electrician tape, tape measure, clippers, fishing line, acetate sheets (for tracing leaves), vials (for bugs), my all-time favorite pooter (a crazy sucking gadget that al-lows safe inhalation of a bug into a vial), camera, Oreos (always!), and water. I tied the bucket to a carabiner, which dangled from a clip on my waistbelt for easy access. I laid an inexpensive six-foot-square, ten-dollar tarp on the forest floor, to minimize mud on the ropes and field gear. We field scientists pride ourselves on ingenuity and an ability to make do. Lab scientists usually require expensive grants to purchase and maintain sophisticated machinery, but field science is typically conducted with relatively modest budgets and a few rough-hewn gadgets.

My fieldwork on that first tree took an entire day to rig ropes and number each leaf on three branches at three heights. It took two more days to do the same thing for trees 2 and 3. I created data sheets as the leaves were marked; each leaf had its own line and twenty-four months of blank squares set up to record monthly changes—herbivory, drying, color change, or death. Once satisfied with the methodology for this activity, I repeated the process for all five species to total three branches at three heights in three trees at three sites of three rain forest types, all within a day's drive. Once the trees were rigged and their leaves marked at different heights, the monthly checks took much less time to revisit. I left nylon blind cords hanging in each tree, so it was easy to hoist the climbing rope into position every month. These slim cords were dark green, innocuous enough to remain invisible and not entangle any creatures. I bought several one-thousand-foot rolls of camouflaged cord, enough to leave in each rigged tree. Fortunately, no lines were ever yanked from a tree crotch. Every month, I inspected each leaf and devised coding to fit on the field data sheets. E stood for young/emerging; G for galls; Y for yellow; and other letters representing different ecological signatures, such as insect frass (aka poop), spider webbing, or presence of caterpillar gangs. Hugh, my loyal colleague from marine biology, volunteered as "dirt" on the ground to record data. Today, an electronic tablet could organize this type of work with far less effort, but we relied on old-fashioned pencils, waterproof paper, and the tedious process of transferring the data onto spreadsheets later. My brain was on overload during those first months of fieldwork and I was exhausted. Words could not express the level of excitement I felt at the forefront of discovery, and it was truly astounding that no one had ever studied the lifespan of an evergreen leaf in the rain forest canopy. Almost every month of sampling inspired a new piece of equipment or adjustment to do things more efficiently and safely.

During the second monthly visit, I was shocked to discover that many of my carefully numbered leaves had been partially eaten. Small holes, big holes, and even galls were evident on most foliage. I was quite affronted anything would dare to compromise those leaf samples, sometimes chewing directly through the ink numbers only recently inscribed. Herbivory emerged as a major factor threatening foliage survival, especially for new leaves. I became adept at recognizing different types of herbivory: chewing, which created outright holes in leaf tissue; mining, which comprised burrowed artistic tunnels between layers of leaf tissue; galls, which were small, swollen infection sites where insects had laid eggs; or fungal attack, which rendered black spots on the foliage. A young leaf is extremely vulnerable to herbivores not only due to soft, chewable tissue, but also because it is oftentimes more digestible before sequestering defensive chemicals. Plant and insects are engaged in an arms race—the plants trying to create foliage more resistant to successive generations of defoliators, while the insects rapidly adapting to digest them. Leaf survival strategies include toxins, toughness, and seasonality of leafing strategically adjusted to avoid insect emergence. Insect herbivores scramble to stay ahead of the leaf defenses, adapt to digest chemicals and chew tougher foliage, hatch out in time to consume new leaves, and overcome any other physical or seasonal mechanisms that plants might develop. For example, a tree that leafs out gradually over long periods of time can minimize onslaught by specific caterpillars that hatch synchronously. Equally important, an herbivore needs to find its foliage of choice amid a vast green expanse. In tropical forests, this includes thousands of different tree species with leaves that are soft versus tough, nutritious versus not, old versus young, common versus rare, already eaten versus still whole, protected by ants versus otherwise, sun versus shade, and high versus low. It occurred to me that the forest canopy was like the most complex salad bar on the planet. Just as humans select from various greens to create a salad, insect herbivores in the treetops confront an enormous choice of vegetarian fare, but only some of which they are adapted to digest. Beetles may fly many miles to find a coachwood tree with young leaves to suit their palate and jaw strength. Leafcutter ants (*Atta* sp.) in Costa Rica may forage several vertical miles up and down many trees to find a supply of a specific texture of *Virola* leaves for their thousands of workers to transport into underground fungal gardens. A specific weevil species may travel the entire length of a sassafras branch to find ideal sun-flecked conditions containing leaves with suitable texture so its proboscis can suck the leaf juices.

Another discovery during those initial climbs into the eighth continent was how leaves changed in size, color, and thickness at different heights. The vertical ascent in most trees was akin to a skyscraper where the penthouse differs significantly from the real estate in the basement—not just in terms of view, but of light, air quality, winds, and just about any other environmental factor. While climbing up the brown, snaky rope (at least to me it looked like a serpent coiling down through the leaf layers), I first encountered understory leaves at lower heights growing in conditions of shade, no wind, and high humidity. Those lower leaves usually slapped me in the face as I climbed because they were large, floppy, and thin. Dark green in color, they were speckled with insect holes and etched with circular, artistic trails of leaf miners. Sometimes the leaves were dirty with pollen or coated by dust due to the absence of high winds or pelting raindrops. Often, they were covered with a layer of moss, lichen, and other miniature biodiversity, things that thrived in the deep shade on a moist leaf surface. This layer of life upon a leaf was collectively called epiphylly, which is a veritable commune of Lilliputian creatures, all living on one phylloplane (a fancy word for the surface of a leaf), and to date only two scientists have dedicated their careers to study it.

Above the understory, the mid-canopy contained mixed leaf sizes and textures—slightly smaller than in the deep shade but larger than in the sun at the top, thinner in shade pockets and thicker when exposed to frequent sun flecks. In the upper crown, leaves became amazingly small, thick, tough, lighter green, and resilient as a result of the hot, dry, windy conditions toward the top. Technically, rain forest trees have two physiological leaf types varying with light environment: shade leaves at the bottom and sun leaves at the top, but with many intermediates throughout the midsection. (To complicate things even more, sun leaves can occasionally be found growing low on a tree at the edge of a gap or a roadside, where direct sun hits lower branches.) The tallest trees, called emergents because they rise above the other trees, received light on all sides, so their foliage was composed almost entirely of sun leaves. Tropical trees, some with an estimated millions of leaves, represent a vast expanse of green, but not a homogeneous green. As a leaf lover for many decades, I can vouch for the complex mosaic of green hues in the canopy—different species, heights, ages, health, light conditions, susceptibility to insect attack, and vulnerability to wind and rain. Like all good real estate, "location, location, location" is the name of the game and greatly determines a leaf's fate. Even chance events, such as nearness to dense arrays of vines or air plants (where insects might hide during the day and feast by night), or distance to bat roosts (which emerge at night to consume hordes of insects in the neighborhood), can impact leaf survival.

Just as the high levels of herbivory haunted me, another anxious thought kept me awake during those first months of sampling: What if insects loved to eat the black ink used to number the leaves? Or conversely, what if they avoided feeding on leaves marked with ink? This type of outcome could create the ultimate sin in field biology—a sampling bias. So I designed a short field experiment to test the ink itself in terms of its impact on insect munching. During a three-month period, I marked one hundred leaves of all ages with a number placed on different regions of the leaf surface. I returned each month and tallied the number of leaves where insects had bitten through the ink versus avoided eating the ink. Thankfully, the results showed no statistical preference by herbivores for consuming, or avoiding, the ink. What I originally thought would be a straightforward thesis to calculate leaf lifespans was quickly becoming a complex array of data.

In addition to the rigors of the data itself, fieldwork continued to be fraught with logistical hazards. During the second year at one subtropical site, I always climbed both giant stinging tree and coachwood, because they grew side by side. One sunny October day, I traipsed into the forest, ropes in hand, enjoying a surround-sound concert of the lyrebird, an elegant Australian ground bird named after its expansive lyre-shaped tail. As part of their beautiful symphony, lyrebirds imitate their environment, usually the calls of other birds. Unfortunately, what I heard instead was a perfect rendition of trucks downshifting gears on a mountain road. I had never heard a lyrebird imitate anything except another bird and felt saddened by this affirmation of human encroachment into Mother Nature's world. That same day, entering a small clearing and looking up to inspect my climbing branches, I suddenly noticed the soil moving underfoot. In the dappled sun shadows wriggled both young and old brown snakes, rejoicing at a new brood of hatchlings and perhaps relishing the sun's warmth on their slippery surfaces. I almost scaled the trunk without a rope, so frightening was this sight of extremely venomous snakes in such large numbers. It turns out some Australian snakes celebrate the spring solstice with elaborate sexual encounters as well as the birth of their young. They engage in a twisting, writhing knot like some sensuous orgy. I reversed my steps in great haste and decided to skip the monthly observation, given the deadliness of a brown snake's bite. In Australia, with over 90 percent of snakes classified as venomous, it is a small consolation that dangling in the trees is safer than standing on the ground—but you still have to get past the snakes and into the trees!

Another logistical hazard of fieldwork was the extensive driving required between patches of remnant rain forest. I added almost a thousand miles per month to the odometers of those university station wagons over three years, from Sydney up the coast to south Queensland and back. On some return drives, I opened all the car windows and sang loudly to music on the radio to stay alert. Kangaroos often jumped onto dark highways only to be blinded by automobile headlights; they caused significant damage to Australian cars, in addition to their own unfortunate mortality. I was lucky—no collisions with 'roos during fieldwork, but that fear kept me awake. The other big danger on the road came from stones sent flying by oncoming trucks.

Australian cars did not have the same requirements for safety glass as those in the United States, so one small windshield crack could become a glassy expanse of shrapnel in no time. I never felt comfortable driving on those rough Australian roads, especially during late hours after exhaustive fieldwork. It was always a relief to get back to my garage apartment in Sydney and lay all my data sheets on the floor to ponder what secrets the leaves had shared each month.

Living on a graduate student budget presented a never-ending logistic challenge. I figured out a few ways to minimize expenses, but in hindsight they were probably quite risky. After field trips, I opened the oven door and used the gas flame to heat the tiny apartment, admittedly not safe but very inexpensive. My living quarters cost only $40 per week but had dangerously outmoded electric wiring and an ancient mattress that probably housed new species of biodiversity in its moldy interior. The front door opened with a generic skeleton key, but I owned nothing of value except data. In contrast to the prior sojourn in Scotland surviving on roadkill, I lived pretty well in Sydney, usually feasting on omelets for dinner, which cost mere pennies. My landlady was ninety-six and invariably pickled on sherry by 4:00 p.m. She was both a watchdog and a "sticky beak," offering her opinion of all male visitors who came along the path to the apartment. I had a revolving guest list since I lived close to the international airport and was frequently asked to meet visitors. Once, I hosted one of the department's few visiting woman scientists, Jean Langenheim, an ecologist from the University of California, Santa Cruz. Jean loved my invention of canvas boots sewn onto khaki pants to minimize leech invasion, so she decided to make her own set. After several hours of hand stitching, she proudly held up the trousers and realized she had sewed the boots on backward. We still laugh about that episode. Sometimes visiting technicians came from other universities. One in particular from the ecology lab in Canberra always brought a favorite bottle of Brown Brothers chardonnay as a house gift, which was just about my only alcohol due to budgetary constraints. When there were no guests, I saved bus fare by riding a bicycle eight miles to the university, departing before traffic in the morning and returning home prior to rush hour. My big splurge was the purchase of a secondhand record player, along with some used 33 rpm vinyl, to listen to music while analyzing data. In the days before iPhones, that contraption filled half the tiny living room!

The simultaneously most frustrating and inspirational part about field research in the rain forest was realizing that every question generated five more. And every answer seemed to create even more mysteries, each of which demanded more climbing. Not just small experiments about ink on leaves, but puzzles about leaf toughness versus insect mouthparts, whether biting or mining foliage was more detrimental to leaf lifespans, and many other unknowns. My notebooks became chockablock with observations, data, illustrations, and thousands of measurements, inspiring multiple doglegs of investigation. I measured leaf lengths, widths, holes, mining patterns, edges, water content, longevity, toxins, thicknesses, surface areas respiration, and even how foliage dried and decayed. Before ropes and harnesses, foresters had studied less than 5 percent of tall tropical rain forests and overlooked the remaining 95 percent where most of the action occurred. Now I was doing the utmost to tackle that other 95 percent. To me, climbing provided access to a frenzied hot spot of unknown creatures and their interactions.

Like all ecology students, I took a statistics course; in my case, I sat through the class twice, once at the beginning of fieldwork and again toward the end, because I desperately wanted to apply the best sampling protocols to my field data sets as a pioneering arbornaut. (Thank you, Tony Underwood, for your incredible lectures illustrating experimental designs using your own intertidal data.) How could I accurately sample leaves in a vast three-dimensional habitat knowing full well I could not measure every leaf? I learned in statistics class that almost everything in nature depended upon two variables: time and space. But I had not really taken these to heart until analyzing my canopy data. When temporal and spatial variation were correctly accounted for in field sampling, it was possible to arrive at accurate conclusions from a subset of leaves without measuring every single leaf. Time: Leaves differed between young and old age classes, and from day to week to month to year. Some Australian rain forest trees had not only seasonal variation of leaves but also physiologically different leaf shapes between seedlings and adults. Space: Leaves also differed along spatial scales—from the top to the bottom of an individual tree or entire hillside, between local stands and entire continents, or in some cases, even micro-locations around one trunk. Evaluating the life cycle of leaves in terms of time and space provided insights not only into leaf survival but could be extrapolated to overall forest health.

Humans tend to make observations over relatively short time spans and usually fail to appreciate the longer-term processes required for some organisms, especially trees. But those two pillars of field biology— time and space—required long-term measurements to ensure accurate data. Monitoring specific leaves over long time frames enabled accurate analyses of change over days, months, years, and even entire forest diebacks over decades. My devotion to spatial variation, achieved through climbing, demystified the complexity of trees from bottom to top. Sadly, it was often difficult to conduct long-term field data collection. Support for research from grants or institutions usually required annual reviews that run counter to long time frames. Scientists are pressured to finalize results after one or at most three years. Conclusions were not reliable when temporal variability was not adequately accounted for. When funders or employers asked me to report findings over a rapid timeline, I collected short-term observations that were essentially snapshots or discrete measurements taken during one time frame. Thanks to many years of data collection in Australia, I was able to calculate the inaccuracy of snapshots as compared to long-term observations. The difference was startling! For example, when I measured coachwood herbivory during one single day, either by clipping a handful of leaves or picking them up off the forest floor as foresters used to do, the average defoliation was approximately 8 percent leaf area loss. But when I measured defoliation over entire leaf lifespans (which amounted to five years for some coachwood leaves), herbivory averaged 22 percent annual leaf area loss. The more accurate result was obtained by monitoring a leaf over its entire lifetime, and exposed a threefold error using the quick-'n'-dirty method of snapshots.

Why this discrepancy? The answer was simple: by picking a handful of leaves and measuring their damage, it was impossible to account for leaves totally eaten by herbivores or to calculate herbivory of the whole tree, including the upper canopy foliage. Only by climbing and measuring over time could I accurately calculate consumption of the entire salad bar.

During PhD research, I took the opportunity to address a hypothesis about leaves in the safe and easily controlled environment of a laboratory. It was a huge contrast to work in a lab, and I enjoyed the small luxuries of wearing clean underwear each day, ordering pizza delivery, and making coffee in my own kitchen. The laboratory question: Which were more toxic—young or old leaves? Some ecologists thought young leaves contained a higher density of chemical toxins that dissipated with leaf expansion. Another school of thought was that leaves accumulated toxins gradually over a lifetime. To answer this, I picked, dried, and analyzed young, middle-aged, mature, and senescent leaves of all five species. They were tested for phenolics and tannins, standard compounds known as defenses from herbivores. In contrast to fieldwork, where Mother Nature ran the show, an analytical laboratory was sterile and controlled, ideal for data collection. But, like most aspects of leaves, the results were not straightforward. One species exhibited its highest toxin levels in young leaves, while three species built up more toxins as leaves aged. A fifth species had almost no toxins at all, illustrating the wide spectrum of defense tactics in rain forest trees.

The phenology of leaf emergence engaged temporal (or seasonal) strategies to protect these green machines, not just to fool hungry herbivores but increasingly to minimize the ravages of drought, floods, storms, and climate change. In many temperate forests, leaf emergence and flowering now occur almost one month earlier in the twenty-first century than a few decades ago, as a response to warmer climate. Because leaves represent the tree's economy, these significant time shifts may well impact whole-forest health. New foliage, soft and fragile like a baby's skin, must strategically avoid predators as well as climatic extremes that threaten its survival. In temperate forests, leaf emergence happens in one springtime burst. While the overall energy efficiency of these solar-powered factories (aka leaves) is only 1 percent, the remaining 99 percent is allocated to the maintenance of the leaf. Half of that 99 percent allocated to maintenance is expended in transpiration, the movement of water from the roots all the way up to the canopy leaves. In addition to sunlight, water is a critical component for foliage productivity. The continuous passage of water up through a trunk to a leaf goes against gravity, and in some trees water may travel as high as 350 feet! Xylem cells transport water collectively and serve as a pump, sucking liquids upward from root hairs to sun leaves. It is estimated a tiny cornstalk can lift 440 pounds of water in a brief growing season, and tall redwoods can transpire 4,400 to 8,800 pounds of water daily! Operating silently without petroleum fuel or toxic wastes, trees vastly outcompete any human-constructed factory.

In those ancient days before mobile handheld devices, I carefully wrote all notes in separate columns on worksheets. Back at the office, I transposed the data for statistical analyses. It was very time-consuming. But boy, did I have data! Each time a bud burst, the numbers of new leaves almost doubled. After eighteen months of climbing, I tallied 4,183 leaves split among five species. Like an addict, I could not stop adding new leaves to the database, and when I wrote up my thesis, the number of original leaves had more than doubled. After thirty-six months of climbing trees, I started grappling with the story unfolding from the data. So, what did I learn? Most rain forest canopy foliage lived three to five years on average, but ranged from a short-lived four to six months for the fast-growing, albeit toxic, giant stinging tree, to six to twelve months for the deciduous red cedar, to over twenty years for understory sassafras leaves. (How did I calculate sassafras, you might ask? To make a long story short, I continued to monitor those remaining leaves for many years after my thesis was completed.) At higher elevations where sassafras trees were smaller and subject to cooler and windier conditions, their leaves only lived two to four years, regardless of height in the canopy. Location, location, location!

Insect herbivory was a primary culprit affecting the fate of a leaf. Before my forays into the canopy, no herbivory measurements had been calculated for entire trees either in Australia or elsewhere, and none with careful replication or consideration of both temporal and spatial variation. Tree foliage offers many different and nutritious types of greenery to herbivores. Even though the restaurant is always open, and trees can't run away from enemies, insects surprisingly never consumed an entire rain forest canopy throughout my lifetime of observations. (But it was different in the dry forests. That story comes later.) For coachwood, herbivores ate approximately 22 percent leaf surface area per year, mostly during the first three months of leaf life, an average of almost one-quarter of every leaf! Holey leaf! Sassafras shade leaves suffered higher herbivory than its sun leaves, but the entire crown averaged 15 percent leaf area defoliation per year. Red cedar had only 4.5 percent annual leaf area loss, consistent throughout the tree. This species did not have well-defined sun or shade leaf populations due to its lollipop structure and short-lived deciduous leaves. Antarctic beech suffered 31 percent herbivory, with a whopping 99 percent of damage in the first three months of a leaf's lifespan when tissue was soft and vulnerable. Stinging trees were the biggest surprise of all. That iconic devil of the rain forest, with its apparent defense of stinging hairs, suffered an average of 42 percent leaf area eaten. This extraordinary percentage just blew me away! One beetle species had evolved to digest the stinging hairs, and no predators ventured close enough to pluck it off those toxic surfaces.

My humble, homemade canopy access toolkit not only allowed me to sample leaves throughout a whole tree, but also revealed that insect defoliation was four or five times higher than prior estimates limited to the forest floor, aka the "big toe" of a tree. For a field biologist, this level of discrepancy was a huge deal! When data collection ended, I was a veritable "leaf doctor." Leaves of the canopy foliage were my "patients." I knew the vital signs of each one: Healthy? Eaten? Intact? Dry? Leaf mining? Yellow? Brown? Torn by wind? Covered with insect frass? Branch broken (by wind or boisterous Australian birds like cockatoos)? Buds? New leaves? Galls? I also knew how to identify damage associated with most types of insects. Fly larvae left distinctive mining signatures on leaf surfaces. Beetles of the family Chrysomelidae tended to take medium-sized bites along the leaf edges, although a few made lacy holes between the venation. Walking sticks boldly ate 75 percent or even 95 percent of a leaf in one sitting. Some Lepidoptera (butterflies, moths) caterpillars or Diptera (flies) larvae ate in gangs, denuding entire branches but leaving the main veins intact because they were too tough to chew. Sucking weevils left brown or yellowish round discoloration splotches on a leaf surface. I was forced to become a self-taught entomologist and figure out what was chewing my beloved leaves. But the illustrated keys, monographs, and books about Australian arthropods were few and far between, so it was more practical to collect and donate specimens to the experts.

On occasion, I took samples to the Australian Museum or to CSIRO (Commonwealth Scientific Industrial and Research Organization), where some of the entomologists identified the specimens and told wonderful tales about the six-legged world. One of the pillars of Australian entomology was a weevil expert named Elwood Zimmerman. I shyly shared my collections with him, and he immediately noted five new species and one new genus. His comments such as "New species of Diabathrarinae" and "Genus? New? In the Imathia group" and "New species of Tychiinae" put a twinkle in his eye. It was obvious to all the Australian Museum curators that the canopy contained many undescribed insects, and now they had an arbornaut to collect treetop specimens. I became a popular figure with these taxonomists, who eagerly accessioned all the critters into their museum drawers.

Just as SCUBA gear inspired coral reef research in the 1950s, my field methods of ropes and harnesses helped launch treetop exploration. Forest canopies became a hot spot for tropical research in the 1980s and 1990s, expanding from Australia to many other tropical countries. Having designed safe access, I started training fellow students both in Australia as well as nearby countries such as Indonesia to become arbornauts like me. Halfway around the world, but coincidentally also in 1979, another graduate student, Don Perry from California State University at Northridge, used a harness, ropes, and crossbow to climb trees in Costa Rica. Like me, Don in turn taught others to climb, and the two of us independently promoted canopy research on different sides of the globe. Five years after our initial climbs, we finally met at a science conference, but only after each publishing about our similar methods in a scientific journal. In the days before internet, science communication was ridiculously slow, and researchers only learned new findings by attending major conferences or reading journals that sometimes took months to arrive in the mail. During those formative years of discovery, Australia was isolated from Central and South America, both geographically as well as academically. At the time of my pioneering forest exploration, Australia reformed her land-use policies, shifting from cutting down rain forests to a massive national effort to conserve the remaining fragments. Generated in part from knowledge learned by whole-tree exploration, the Australian media and policy makers reversed their outmoded mindset, becoming a voice for conservation instead of deforestation. I am proud that my canopy discoveries helped generate this transformation. Today, Australia's small remaining tracts of safeguarded rain forests are valued for both ecotourism and their important multimillion dollars' worth of biological services: soil conservation, carbon storage, productivity, biodiversity, fresh water, and climate control.

At the time of these pioneering treetop studies, my marine biology colleagues were similarly engaged in trying to understand the functions of their high-diversity coral reef ecosystems. We were all intoxicated with discovery, and sleepless at the notion of figuring out how so many species lived in one place. Both rain forest canopies and coral reefs were considered essentially the cutting edge of ecological field research. But during the 1980s, we were oblivious to the notion that the climate was changing and how this might impact our data collection. In 1982, the world's strongest El Niño was recorded and the eastern Pacific waters became very warm, followed by a first major coral bleaching event, extending to Panama, Central America, and the Galapagos. In the Panama Pacific reefs, 70 to 90 percent of all corals died. Most coral reef ecologists called this a "coral problem" until it recurred around the equator in 1997–1998, at which time the international scientific community took note. Similarly, rain forest ecologists considered logging, burning, and clearing as the major obstacles to forest health in the latter twentieth century, with relatively few discussions about climate change until the early twenty-first century. While I was dodging logging trucks in the 1980s, I did not sound the alarm about some of the astounding insect outbreaks and their associated warmer, drier conditions because we ecologists thought these episodes were simply natural cycles, with a modicum of human intervention. Both coral reef and rain forest scientists were still focused on learning how these highly diverse systems operated and were not really paying attention to the enormity of human-created aberrations, especially climate change, whose threat was accelerating.

❧ Coachwood ❧
(*Ceratopetalum apetalum*)

WHEN I SAW MY FIRST COACHWOOD, it was love at first sight. What arbornaut wouldn't aspire to spend time in the boughs of this sturdy, beautifully apportioned, white-trunked, elegantly branched tree with brilliant green leaves and flowers that turn from white to pink as they age? It was an ideal species for climbing, discovery, and biodiversity. The common names of scented satinwood or coachwood refer to its fragrant bark and useful timber for constructing coaches. Its scientific name translates to *ceras* for "horned" and *petalum* for "petals" because of its hornlike lobed flower petals, and the species name *apetalum* means "no petals." Coachwood's abundant white flowers turn pink after budburst, but the coloration is due to petal-like sepals, not true petals, hence the term *apetalum*. The flowers provide shelter and food for a large variety of invertebrates, as do the red fruits, which attract king parrots and other frugivorous birds. Coachwood belongs to the family Cunoniaceae, boasting some diverse tree cousins such as *Vesselowskya*, *Pseudoweinmannia*, and *Schizomeria* (all mouthfuls to pronounce as well as tough for any nonexpert botanist to identify in the field since they look similar).

Because coachwood was my first climbing tree in the rain forest, I was especially attentive to the diversity of life in its upper reaches, living amid those varied and economically different types of leaves. A few of its residents were particularly memorable: long-snouted black weevils (*Apion* sp.) sucked the juices of understory and mid-canopy foliage; a powerful owl (*Ninox strenua*) roosted in its branches for a daytime siesta; flocks of parrots noisily frolicked in the upper branches to eat fruits; and *Edusella* sp. and *Colaspoides* sp. beetles preferred young leaves (but not old). As I always advise students, most discoveries lead to more questions than answers. Coachwood canopies still have many secrets to reveal to us ground-limited bipedal beings.

Coachwood is common in Australia's warm temperate rain forests throughout New South Wales and southern Queensland, growing in relatively poor soils, including the gullies of urban Sydney bushland. As a slow-growing tree, it produces strong timber and lives up to two hundred years. Its brilliant green foliage has very inconspicuous toothed edges, opposite in configuration, and is delicious to a range of herbivores. In the uppermost crown, coachwood sun leaves exhibit one of the highest recorded toughness values of any Australian rain forest species, probably to minimize drying out in such an extremely hot, sunny location. My research revealed that, like other rain forest species, coachwood leaves suffered highest herbivory during the first three months in a leaf's life. Shade leaves lost approximately 35 percent leaf area per year to herbivores, while sun leaves lost only 9.4 percent. Similarly, coachwood leaf decay required twenty-one months to decompose on the forest floor, typical of a tough, sclerophyllous leaf. Its leaves were the toughest of all five species I studied, but contained only a moderate level of phenolics, implying their hard, waxy surface was more important than chemical toxins for repelling herbivores. (In contrast, stinging trees had almost no phenolics, and mature sassafras leaves and Antarctic beech had almost twice as many.) I studied coachwood because its leaf was typical of many other Australian rain forest trees: sclerophyllous (meaning tough cuticle), evergreen, nearly entire (but slightly toothed), elliptical, and with reddish young leaf coloration. Whereas temperate foliage often turns red or orange in autumn before leaf fall, many tropical trees produce red or pink pigments during leaf emergence. This flash of color may be Mother Nature's way of creating a deterrent for herbivores since pink fools any herbivores that might otherwise be searching for green as the preferred food color. It may also be an economic strategy for the tree, to avoid an early investment in the relatively expensive green pigment, chlorophyll, since young leaves have a greater chance of being eaten.

I nicknamed coachwood my "economic tree" because it taught me so much about the business plan of a healthy canopy. Whereas humans invest in stocks, real estate, and furnishings and sometimes measure success by yachts or wine cellars, a tree has a similar blueprint for survival. A deciduous tree follows a pretty clear-cut business plan of shedding its assets every fall, with blind faith that sunlight and warm temperatures will ensure the onset of new foliage the next spring. These tiny units of green represent machines that collectively impart growth to the tree, bringing in sugars and allowing expansion to outcompete its neighbors. As an evergreen subtropical crown, coachwood has a much more complex business plan, where each leaf has its own unique formula for success, representing a fairly complex architecture of diverse units. The lower leaves are larger and darker to efficiently utilize the tiny amounts of light that filter down through the upper crowns; these leaves also live a long time, because it is expensive to produce them. In contrast, the economics of the upper foliage centers around much smaller, thicker, and yellowish-green leaves, perfectly suited to the hot, dry conditions, and they comprise extremely high-powered chlorophyll factories to produce sugars that keep the entire tree alive, healthy, and growing. In between the sun and shade leaves were variations of each, all built to suit a specific environment in the complex canopy system. Coachwood sun leaves are approximately one-quarter the size of the understory foliage, as well as different in thickness, color, toughness, moisture, and function. Canopy growth is essential—all tropical trees need to produce a crown that outcompetes their neighbors for light, plus a root system that maximizes intake of water and nutrients. Each individual possesses a blueprint that intricately coordinates when every leaf emerges, becomes net productive, maintains itself, and eventually falls. If a branch gets knocked off in a storm, it's a serious injury: the entire crown must immediately recalibrate, adjusting for the altered architecture and also recalibrating the function of each leaf in the neighborhood, similar to infrastructure adjustments made after an earthquake disrupts a city. Thanks to *Ceratopetalum apetalum*, with its complex salad bar at different heights and light levels, I have a greater admiration of the complex machinery that comprises a rain forest tree.

4

WHO ATE MY LEAVES?
≫≪

Tracking—and Discovering!—Australian Insects

AFTER A YEAR OF FINDING HOLES in leaves throughout every canopy, I was on a mission to track down the insect marauders whose jaws were tunneling extensively through all my Antarctic beech leaf samples. Dangling from a rope at eighty feet in the tree crowns, I was queasy from the height but also from the damage I saw around me—nearly every new leaf of these several-thousand-year-old trees was riddled with brown tunnels or tiny holes. The lines of silk I found myself swatting as I moved through the air provided a clue that these leaf eaters were larvae, literally wafting on their threads from branch to branch. It looked like only the young leaves were being eaten, so I anticipated my miscreant was a very tiny (translation: hard-to-find) critter whose small mouthparts could only eat the softest vegetation. Into the upper reaches of beech, I started to notice tiny wriggling white blobs, less than half an inch long, on every branch. These squiggly larvae were tunneling through one layer of tissue on the youngest leaves at the branch tips. As they grew bigger, they moved farther down the crown to eat slightly bigger bites of tougher, middle-aged leaves. These herbivores graduated to older foliage as their jaws became larger and stronger.

I was not a trained entomologist and had no idea how long it might take the larvae to metamorphose, but knew that I needed to hatch them into adults in order to identify the species. So I carefully transported a few dozen plump larvae, plus a carload of beech branches, back to my garage apartment in Sydney. Using buckets to hold the beech branches surrounded by plastic bags to keep the larvae from ballooning off on a silk thread, I patiently reared them in the small confines of my student living room. After about ten days of feeding on young leaves followed by slightly older foliage, the tiny white blobs had grown fat and squishy although still only about a half inch long, with legs on both ends but not in the middle (an identifying characteristic of Coleoptera, the order of beetles). One day, they started dropping from their beech branches to the floor, getting lost in the fifties-style green shag. They had almost magically transformed from larvae into spherical crystals, which is the beetle equivalent of a moth's cocoon. Panicked (because I did not want to step on them!), I was able to carefully sift through the carpet threads and extract each pupated sphere that represented the next stage of beetle metamorphosis. Like a mom giving birth, I was giddy with joy observing these larvae pass through several instars (aka stages), then emerge a few weeks later as handsome cinnamon-brown beetles.

I dashed off to the Australian Museum with a few vials of both pickled larvae and adult beetles, excited to share this discovery with some entomologists. The insect curators all clustered around and then proclaimed these critters were chrysomelids, a specific family of beetles. The world expert, Brian Selman, resided in England, so we mailed a few specimens to him at the University of Newcastle upon Tyne. Museums have legal permits to exchange specimens, so my precious vials were safely transported halfway around the world. Despite having never climbed a beech himself, Brian was able to write up the technical description of this new host-specific herbivore using my field notes, and I was his proud coauthor. Initially, Brian was frustrated because he had just completed a monograph on chrysomelids and now he had to revise his lengthy publication to include this new addition. But taxonomists are always revising, editing, and adding to their evolutionary trees of life as species are discovered and/or classifications change. He named the new beetle *Novocastria nothofagi*, in honor of his institution (Newcastle) and the beetle's host tree, Antarctic beech (*Nothofagus moorei*). We never met in person but created a professional bond through our global passion for beetles. Like so many insects living in the eighth continent, it was not only a new species for us, but also a new genus. This means an entire evolutionary group of beetles (a genus henceforth named *Novocastria*) had been made known to science. Many insects don't have common names, just scientific ones, but I affectionately nicknamed this one the "Gul beetle," to honor Williams College, because *gulielmensian* is Latin for "Williamsian," and also the name of the school yearbook. Finding, classifying, and publishing new species is a painstakingly slow process—over two years in the case of this new beetle. Arbornauts estimate that 90 percent of canopy species remain undiscovered and unclassified. Over the years, as critters have been discovered in forest canopies, they have traveled to entomologists worldwide: to British coleopterists, Russian mite experts, Australian weevil curators, Smithsonian Institution arachnologists, Floridian bark beetle laboratories, and others in between. Like the treetop ecosystem, the world of science is composed of a vast web of experts who collaborate on different facets of natural history.

I did not set out to study insects in the treetops, only leaves, but I became frustrated and puzzled to find recurrent signs of insect damage every month of fieldwork. Most herbivores left big holes from chewing, but others left gorgeous, almost artistic designs from tunneling through layers of tissue, usually leaving the cuticle intact but burrowing through the softer moist interior of the leaf. A few insects created braille-like messages in the form of galls (where midges or wasps lay eggs inside leaf tissue and the activity of the grub causes a swollen nodule to form). Sometimes, additional clues were left by the marauders in the form of small black pellets, called frass, which indicated an insect had been feeding quite recently and the excrement had not washed off the leaf surface. Had today's DNA techniques existed back then, I probably could have analyzed the poop to identify the culprits. Instead, patient observations were required to witness the diabolical

act of an insect chewing on its lunch. Having discovered all this carnage, I needed to become an insect detective if I was going to continue
my work as a leaf specialist: Who ate what? When? Where? And how
much impact did it have on these trees?

Almost all plant species undergo defoliation because leaves cannot run away from their enemies. Mobile organisms can quickly move
away to escape predators, but plants are rooted in place. So they develop
sophisticated strategies to protect themselves: they produce toxins to
render their tissue poisonous or unpalatable, or escape using time (by
leafing all at once, so some leaves escape attack) or space (by "hiding"
in a forest surrounded by other tree species). Trees are especially vulnerable to being eaten due to their lack of mobility and the fact they
have invested decades of growth into becoming a mature tree. My detective work to find the grazers was challenging given their tiny size
amid an enormous salad bar of greenery. Large herbivores such as
deer, giraffes, elephants, or sloths on other continents are easily visible,
whereas Australian rain forests have none of those bigger vegetarians
(except koalas, which only live in dry forests where they exclusively
eat eucalypt foliage). Insects created a whole new level of complexity
and confusion to my field schedule as I struggled to deploy new methods and more time in the trees to demystify the interactions between
leaves and what munched them.

Finding an insect in the forest canopy seemed akin to locating a
needle in a haystack, and I was discouraged. I had successfully found
one important herbivore in the Antarctic beech canopy but knew that
almost every other tree species had one (or more) leaf feeders. How
could I design fieldwork to discover some of the most important marauders? Most munching was a cryptic event, with insects both camouflaged on the leaf surfaces and feeding alone, making them really hard
to find. Based on ecological literature, I knew insects were actually
incredibly abundant in vegetation, yet elusive. Maybe it was like birdwatching, where years of practice leads you to develop an eye for the
quick flutter of wings. Instead of spying on feathered flyers, I needed to
learn to spot small, scurrying feeders on leaf surfaces. "Biodiversity" is
the collective term describing the variety of species on Earth, of which

many are six-legged, aka arthropods. In the 1800s, Charles Darwin estimated that Earth housed eight hundred thousand species. (I can only imagine the queen of England gasped and was most impressed with what seemed an enormous number at that time, as calculated by her prodigal young naturalist!) Nearly one hundred years later, a scientist at the Smithsonian Institution named Terry Erwin increased Darwin's original estimate almost thirtyfold, just based on beetles he counted living in a tropical tree. Erwin sprayed pesticide overhead using a handheld mister to bring the arthropods down to the forest floor, where he counted the insect "rain" that fell onto a plastic sheet spread out under a tree canopy. From his fogging episode, Erwin extrapolated that the world may contain approximately thirty million species, and the majority were insects that had not previously been classified, a healthy majority of which were beetles. And because over half of all insects are reputed to be herbivores, ecologists now estimate that over 50 percent of the world's terrestrial biodiversity lives in the canopy with its gazillions of leaves. In 1988, Professor Edward O. Wilson, the distinguished entomologist at Harvard University, upped Erwin's estimates and speculated that Earth contains over a hundred million species, including bacteria and soil organisms. Recognizing that scientists have not adequately explored either the treetops or the soil ecosystems, Wilson believes both regions contain many undiscovered species with the potential to raise his estimate further still. But the speed of discovery of new species in rain forests, as I soon discovered, is extremely slow due to that "needle in the haystack" phenomenon, so scientists estimate that approximately 90 percent remain undiscovered; hence the term "eighth continent," coined independently by several of us arbornauts, reflects the undiscovered expanse of forest canopies. Many of those undiscovered species may disappear without humans ever knowing. Even more mind-boggling is the number of insects living on our planet, estimated at ten quintillion (written as 10 plus eighteen zeroes). Of this total, a huge proportion of those six-legged critters live in the eighth continent.

Once I started focusing on the millions of unknown insects in tree crowns, I also started to appreciate the multitude of insects in our daily lives. It turns out insects are abundant yet invisible in almost every ecosystem, not just the eighth continent. How many of us know the number of arthropods that inhabit a cubic yard of our own backyards? Or the amazing creepy-crawlies living in our own homes with us? And most definitely, we have no idea how many insects live in forests, except to admit it is a big number. Science has advanced with so many astounding calculations: the distance to the moon, the diameter of an atom, the dimensions of a dinosaur, and even a map of the human genome. But we have not yet accurately counted the number of insects in a tree crown. Exploration of the eighth continent lags behind coral reefs, deserts, polar regions, and even outer space because only a handful of professional arbornauts exist, so we must scramble to catch up.

Once I realized insect herbivores were a huge threat to foliage and their feeding activities were difficult to observe, I needed to figure out some new methods to up my game as an amateur entomologist. My monthly field trips to observe leaves and chronicle their fate expanded to include a bug component. I had to sleuth out their consumption: who, when, how much, and how often. I added vials and insect nets to my field gear. In addition to swinging in a harness to sample leaves and then insects, I developed a pretty regular field-trip routine: pack, drive, sleep, wake up, climb trees, measure, photograph, catch insects during daylight, climb again, measure, monitor, shower (cold only), dinner, chase resident quolls (*Dasyurus viverrinus*, a marsupial relative of the Tasmanian devil that eats both insects and small animals) from stealing my dinner off the barbecue, sample insects by dark, collapse into sleeping bag, and fervently hope no one stumbles into my remote campsite after a long, liquid night at one of the rural pubs.

During the first year of fieldwork, I learned the skills of a good detective spotting clues of herbivores, ranging from frass to feeding signatures on a leaf. During my second year, I became woefully aware that herbivory resulted in significantly less chlorophyll, reducing the tree's capacity to photosynthesize. After solving the mystery of the Antarctic beech host-specific beetle defoliator, I puzzled over an ongoing episode of coachwood defoliation in the subtropical crowns at the Dorrigo National Park field site. Almost every new leaf suffered some level of damage, but I rarely observed even one insect in the act of chewing.

Then, during a routine monthly sampling trip, serendipity struck. I always camped in a tiny expedition tent because no hotels existed nearby except for a few rough pubs. The camping area in Dorrigo was called Never Never, an apt description because there was never anyone else at this isolated spot. I usually had both the outhouse and lone picnic table to myself, shared occasionally with a few bowerbirds and curious parrots. Male bowerbirds built courtship structures from sticks on the forest floor and decorated them with blue fruits and flowers to woo a female. Sadly, many of the bowers at Dorrigo were cluttered with blue drinking straws, litter from fast-food outlets. I always felt scared when a car drove into the parking area, but my biggest fear was failing to find an insect herbivore or losing track of a marked branch along the vertical rope transect. One starry summer night in February, I awoke at about 2:00 a.m. to use the outhouse. My footsteps crackled on dry leaves underfoot as the only other sound in the forest. Pausing to appreciate true darkness in this remote place, I heard some raucous, grinding noises overhead that reminded me of a truck shifting gears. In the woods, that was downright scary. I returned to the tent for a flashlight and entertained visions from that favorite childhood fairy tale of Jack up his proverbial beanstalk. As I aimed a narrow beam of light into the greenery above, to my amazement, thousands of beetles were munching on coachwood leaves, their metallic carapaces reflecting in my flashlight beam. Eureka! The aerial salad bar was frequented by nocturnal insects. It made sense for most insects to feed at night and avoid predation by birds during the day. I had been searching during daylight hours, which explained why I was coming up empty-handed! This was an exciting discovery, and subsequently useful when future entomologists started to explore forest canopies in other parts of the world. Thanks to my bladder, I was one step closer to demystifying the complex interactions between insects and foliage.

From that day forward, I included night climbs as part of fieldwork. It was spooky to ascend alone in the dark, and I had to exercise utmost caution to minimize encounters with anything poisonous, toxic, or aggressive. There was no exact precaution for avoiding potential predators while climbing a tree in the dark, but careful advance surveillance with my headlamp alerted me to any eyeballs overhead (usually spiders), dangling "strings" of nasty wait-a-while vines (if you got hooked, they never let go due to backward-gripping spines), and occasional brown bumps, which were marsupial tree possums. What a special privilege to share a whole new world of nocturnal vegetarians, not only chrysomelid beetles as in the beech canopies, but also katydids, stick insects, caterpillars, and sucking weevils! The landscape of a leaf surface, aka the phylloplane, was ever changing. Insect feeding frenzies were in full force in the treetops, and I was one of the first to climb up there to observe them in situ!

Not all insects escape their enemies via darkness. Some use temporal strategies, undergoing synchronous hatching to devour leaves en masse during a short period of time before predators detect and then eat them all, or before a tree mounts its own defense by producing defensive chemicals. The Antarctic beech beetle followed this strategy. Beech composed 95 percent of the cool temperate canopy (hence termed monodominant), and one insect, *Novocastria nothofagi*, survived by engaging a feeding frenzy on young beech leaves followed by rapid metamorphosis. Its life cycle was a race against time: enemies such as birds or infections could not build up quickly enough to control the explosion of larvae in the victimized beech crowns. In contrast to beech beetles, other foliage feeders survived via a strategy of rarity. If you are a lone walking stick in a sassafras tree, predators probably won't spot you. Or if you are a katydid feeding solo at night, you have a double insurance policy against discovery by your enemies: escape in space via isolation, plus the cloak of darkness.

Like most field biologists, I always worried about the accuracy of my field methods. One of the fears in designing research is the challenge of avoiding bias. Finding an insect allowed for no ambiguity: either it was counted or it didn't exist. We simply call this "presence or absence." Calculating defoliation requires multiple samples to create averages and thereby achieve objective data; it was especially challenging to obtain accurate estimates throughout the enormity of a three-dimensional canopy, as compared to the controlled conditions of a laboratory. It might be tempting to select leaves that were not eaten, reducing time and effort to calculate defoliation, but that would result in a biased sample. So how could I be sure that my leaf samples and herbivory calculations were an accurate reflection of the entire canopy? Ecologists often use a simple table of random numbers to generate unbiased selections of whatever needs subsampling. For example, if you want to sample ten leaves out of thirty, you need to number them all and then consult a random number table to generate your subsample selection. The purpose of subsampling is to save time and energy (you avoid measuring every single leaf in the forest) and to avoid bias such as picking the closest, prettiest, or least eaten leaves. I learned about the importance of avoiding bias from my statistics professor, who told us about an expensive laboratory trial to study the swimming speed of an important oceanic fish. Every month, the scientists picked thirty fish out of a large tank of three hundred individuals and measured how fast they swam in a special chamber. After two years and great expense, the experiment was canned. By simply plucking fish out of the tank, instead of numbering them and then selecting random numbers from a table, they had inadvertently chosen the slowest-swimming fishes— the ones easiest to catch. (A nightmare for those ichthyologists, but their story will stick with me for life!)

To accurately sample insect consumption in a tall tree, I needed to understand a whole lot about holes in leaves. What was the range of defoliation for different tree species? What age of leaf tissue was preferred by herbivores? Third, and perhaps most critical to all leaf detective work, how could I accurately calculate leaf damage in a whole tree, many made up of millions of leaves, without harvesting the leaves or measuring every single one? These types of sampling questions burned in my brain, and unfortunately no prior publications offered protocols. Another big leaf-worry literally kept me up at night: If a bug takes one tiny bite out of a young leaf, does the resulting hole expand when the leaf becomes full-grown? In other words, does 10 percent chewed on an emerging leaf remain 10 percent when it matures? To answer the "holey leaf" conundrum, I designed a simple field experiment to compare sizes of holes in young versus full-grown foliage. The equipment list was amazingly inexpensive: two-dollar paper punch, waterproof Magic Markers, tags for branches (to find them again each month), and a notebook. I found a treefall that created a small light gap where understory coachwood branches took full advantage of sun flecks. Due to the high light levels in the clearing, numerous new leaves flushed close to the ground instead of ninety feet overhead. I numbered nine to fifteen leaves on each of three branches of three trees, carefully including four age classes: young, mid-aged, mature, and senescent. I used the paper punch to clip 0.33 cm (exact area of one punch), 0.66 cm (two punches), or 1.0 cm (three punches) from random leaves, and placed the holes as beetles would chew, avoiding major veins, which insect mouthparts can't easily bite.

I punched all the leaves and waited. Patience is a requirement for most ecological research. Some data collection takes years, even decades, to complete. In this case, it only took several months for all the foliage to reach maturity. I was excited to return and harvest the samples, all carefully labeled. Back in the botany laboratory, I used a digitizer (a gadget that measures the surface area of any two-dimensional object placed on a belt running over a laser beam) to calculate if the holes had grown, shrunk, or something in between. I remeasured each punched hole in the young leaves that had expanded, in the old leaves serving as controls, and in the mid-sized leaves that underwent moderate growth. For every sample, the holes grew proportionally larger as the surface expanded but remained the same percentage of total leaf area. This was good news—if 10 percent of a young leaf was eaten, it remained 10 percent as an adult. It was a relief to know I didn't need to worry about sampling error when measuring herbivory on a young leaf versus mature; the proportional amount of damage remained consistent regardless of age. This also meant that expressing data as percentages was more accurate than calculating millimeters of holes. Ten square millimeters (0.4 square inch) of young leaf damage turned into forty square millimeters (1.6 square inches) when it quadrupled in size.

Another challenge that plagued me: How could I account for leaves that were 100 percent eaten? If an insect eats an entire leaf, there is no easy way to find and measure it. Fortunately, the dogged process of monthly observations over several years allowed me to determine exactly how many were entirely eaten. Insects usually left clues after eating a whole leaf, such as frass, silk, or a dangling petiole. If I had a leaf numbered 8 located between leaves 7 and 9, but then it was replaced by a pile of frass on the stem—voilà—number 8 had likely been eaten. Usually, petioles remained intact because they were too tough to bite. Typically, I found foliage half eaten in month one, three-quarters eaten in month two, and then completely devoured by month three. High winds also contributed to entire leaf removal, but it was easy to detect damaged foliage due to storms because the entire canopy was impacted. My long-term data set revealed that trees suffered three to four times higher defoliation than previously estimated by forest scientists who made quick surveys limited to the forest floor. Earlier forestry publications cited that forests incur 5 to 8 percent annual defoliation, but my results proved that trees sustain much higher damage. Canopies tolerate insect attack ranging from 15 to 25 percent annual leaf area losses, information that will aid conservation and forest management in modeling healthy forests and insect outbreaks. With the onset of climate change where insect outbreaks are predicted to be on the rise, it is important to gauge the threshold of tree resilience.

The material ingested by insects quickly recycles back into the soil via frass. Rain filtering through the canopy, called throughfall, washes the pellets to the forest floor where they are rapidly reabsorbed by the root hairs. This is an important pathway of nutrient cycling. Conversely, when uneaten leaves eventually fall to the ground, they undergo more gradual decomposition due to their large size and waxy surface. The foliage of four out of five of my research species required more than a year to decay on the forest floor. (Yes, I carefully measured their decay rates, by placing thirty mesh bags containing equal weights of leaf material on the forest floor, harvesting three each month for ten months to weigh them, and thereby tracking the decay rates!) Insect defoliation may be beneficial to a tree because the digested tissue falling to the ground as frass rapidly recycles the nutrients. Severe droughts may circumvent this presumed advantage of frass reabsorption on the forest floor if the soil dries out and surface root hairs die. But severe droughts and their associated heat waves also lead to more insect outbreaks and ultimately more frass. These pathways are indeed complex.

Unfortunately, unlike birds or fish, trees can't shift their location when conditions become unfavorable. At some point, forests exceed a tipping point of stress from the trio of drought, warming, and insect outbreaks, and they die. In the past few decades, insect attacks have surged with hotter, drier conditions created by human-induced climate change. Throughout the 1980s, the topic of climate change was mostly limited to geoscience and climatological circles, rarely making its way to ecological professionals. It is hard to believe that, only thirty years ago, many disciplines lacked that perspective about global change. We were so consumed with figuring out how ecosystems functioned and how so many creatures coexisted that we did not connect the dots and interpret the climate warning signs. Yes, we missed the forest for the trees. Had tropical ecologists recognized the significance of global warming sooner, perhaps we would have conducted biodiversity surveys throughout different forest ecosystems, to establish baseline data before climatic extremes began jeopardizing species survival. Now scientists are scrambling to catch up, as many forests burn out of control or insect epidemics increase in severity and frequency. We cannot know how much is extinct if we never knew what lived in these forests in the first place.

In addition to the potential biases of sampling leaves throughout a vast canopy and the challenge of calculating herbivory of leaves totally eaten, a third dilemma of fieldwork involved human bias. What if some people see things differently than others due to poor vision or tendencies to exaggerate? One solution is to engage multiple samplers to collect data. Believe it or not, one person might be overly timid in sweeping insects into a net, underestimating the counts, thus creating bias—a big no-no. I learned how to minimize human error by engaging teams of volunteers, now known as citizen scientists, to wield nets and count insects. Over time, a few hefty, macho net swingers who broke every branch in their path offset the one or two timid introverts who captured almost nothing in their delicate swings. So, how did I coerce fifty people to enter the jungle and sample insects, you might wonder? Earthwatch! This innovative organization, headquartered in Boston, serves as a clearinghouse to match volunteers with scientific research expeditions. For my first-ever grant application, I applied for an Earthwatch grant, requesting volunteers to measure herbivory in Australian forests. Grants are the lifeblood of science; almost all research requires outside funding to pay for equipment, staff, travel, and even such esoteric things as snakebite kits or laboratory safety goggles. The funding for grants is highly competitive, and newcomers are usually considered less qualified than the old guard, making it tough to break into the system. Carefully answering all the questions, I submitted the proposal and waited anxiously. Several months later, I received a YES letter. Oh joy! Getting a grant is one of the biggest drivers of happiness (and success) for scientists. Getting a NO is one of the worst experiences, but it happens to everyone because most grant applications have only a 5 or 10 percent chance of success. After three decades of field research, I have been fortunate to obtain millions of dollars of grants, but I have been turned down for an almost equal amount.

Earthwatch grants were small in dollars, but generous in human power. Teams of fifty eyes searched in the trees simultaneously, instead of just my two (sometimes very tired) eyes. For eight years, over 250 citizen scientists from many countries contributed to my canopy research in the Australian rain forest. They were all curious, intrepid, and enthusiastic. I persuaded one of the university lab technicians, Wayne, to assist on the expeditions. A former military man, Wayne created elaborate happy hours in the middle of the Queensland rain forests—evidently, he had displayed similar bartending feats for his platoon. He was also a wicked aim with the slingshot, so we became a dynamic duo in the Australian jungle. On one expedition, a volunteer who was a former air force pilot got her hair stuck in the metal ascending device while eighty feet up a rope. We hoisted up some scissors so she could extract herself and prayed she would cut her hair and not the rope. The rescue mission was successful, and to this day, she remains a lifelong friend. Another eager volunteer was a hunter from Kansas. He ridiculed my homemade slingshot, then went home and ordered me a sophisticated one from an American hunting supply company.

Such mail-order weaponry did not exist in Australia, and slingshots required permits. Ironically, guns could be purchased in almost any country store, due to their widespread use by farmers to control vermin. Several months after that expedition, I opened my farmhouse door to find a police officer. He explained I had received an illegal weapon from overseas that required a license. I willingly filled out the paperwork, mailed it in, and became the proud operator of a soon-to-arrive sleek slingshot. When the package was finally delivered, I eagerly opened it. To my surprise, it not only contained a stunning aluminum wrist rocket, but also two pairs of camouflage bikini underwear and an elegant bottle of perfume. A small note at the bottom said, "Knickers and perfume are for Wayne's fiancée." Wayne had become engaged during the last Earthwatch expedition. I can only imagine the customs officials had a good laugh while inspecting that package, wondering what kind of kinky science required perfume, lingerie, and a slingshot.

After that volunteer's hair was stuck in her climbing hardware, the owners of the ecotourist lodge where the teams stayed voiced some concerns. What if someone fell? Were the branches safe? I had to agree it was risky to train so many novices. One evening, over a bottle of good Australian wine, the lodge owner and I sketched a "treetop trail" on a paper napkin. Why not build an aerial path, so volunteers did not need to swing around on ropes? In true Aussie style, we celebrated our idea with more wine. I was sleepless with excitement. An elevated walk would allow multiple researchers to collect data all at one time, regardless of weather or darkness, and allow access to problematic species like the giant stinger, unsafe to climb due to its weak wood and toxic spines. The following year, thanks to some local engineers and lodge employees, the world's first canopy walkway was constructed in Lamington National Park, in southern Queensland. My Earthwatch teams were the first researchers to use this fabulous structure. It was built on a slope, with telephone poles supporting the bridges. That walkway is now over thirty-five years old, and has transported thousands of visitors into the treetops, transforming public perception of jungles. Globally, aerial trails have catapulted canopy research to new heights.

Soon after, a second walkway was constructed at Lambir Hills National Park, Malaysia, built by an engineer who encircled tree trunks with rubber necklaces, allowing bridges to span between tall trees, higher than the limits of telephone poles. After many years of aiming slingshots and hauling ropes over high branches, arbornauts can now use walkways for whole-forest access with greater safety. Today, skywalks are scattered in over fifty forests, constructed with either pole structures or tree trunk necklaces, and made of timber, aluminum, or steel. I often consult on placement to maximize education about biodiversity, working with engineers and arborists who provide the construction skills. I have helped design and build walkways around the world, including the longest tropical structure in the Peruvian Amazon; the most expensive one ever constructed (per foot) in the tropical forest of Biosphere 2 (a huge experimental glass dome in Arizona); North America's first public skywalk, in Myakka River State Park, Florida; the world's first ribbon bridge (made from cement) on Penang Hill, Malaysia; and the most kid-friendly walkway featuring a rope spiderweb, at Quechee, Vermont. In 2020, I launched a new project called Mission Green, with a goal of building aerial walks in the world's tallest biodiversity forests that do not have such access. If successful, this program will provide local income to indigenous people from ecotourism instead of logging, and ensure forest conservation through local stewardship. (More about this conservation effort later!)

But the Queensland aerial trail was the first. Ropes and harnesses were great for solo efforts, but not for teams of researchers. A bottle of wine and a serendipitous climbing incident inspired innovation, creating a new canopy tool that ultimately advanced forest conservation. Walkways remain my favorite research tool due to safe access and inclusivity, including use by wheelchair-bound students. These structures also allowed me to tackle a mystery even more puzzling than those voracious beech and coachwood herbivores: What crazy critters had the audacity to eat Australia's most toxic tree, the giant stinger (*Dendrocnide excelsa*)? I was fascinated by these trees because it seemed nothing should be capable of eating their foliage, defended by both physical and chemical stinging hairs. I could not even climb this species without getting "stung." Instead, I climbed any adjacent tree with a supportive branch that allowed access to the stinger canopy and wore leather gloves for protection while cataloging its leaves. The advent of walkways allowed close inspection of these nasty leaves without risk of brushing against their toxic surfaces. And the results were totally unexpected: more than 40 percent of a stinging tree's leaves are eaten on an annual basis! The responsible herbivore's feeding patterns looked like my great-aunt's antique lace doilies. What the heck was going on? Because all the leaves looked so similar in their damage patterns, I hypothesized that only one herbivore had adapted to digest the toxins and determined to find it.

The giant stinging tree (*Dendrocnide excelsa*) is a cousin of the small nettle that grew along the upstate New York roadsides of my childhood. Its botanical family, Urticaceae, is composed of 2,625 species, most of which grow in the Asian tropics. Tropical forests of North Queensland hosted the species with highest toxicity, called Gympie-Gympie (*Dendrocnide moroides*), whereas its cousin, the giant stinger (*D. excelsa*), grew slightly farther from the equator in subtropical rain forests. These trees were planted in trenches during the Vietnam War as a physical defense against attack. In Australia, a few local pub yarns circulated about an occasional tourist selecting the gorgeous emerald-green leaves as a substitute for loo paper in the bush. The notion of rubbing your backside with those toxic hairs invariably caused howls of laughter, but left folks squirming on their barstools. I always told my volunteers that story, so they would not make a similar mistake. After months of searching, I located the giant stinging tree beetle high up in the crown, resting on the tops of leaves, a position tough to detect when looking upward. It was a well-camouflaged, brilliant green chrysomelid beetle (*Hoplostines viridipenis*), and the lacey nature of its feeding pattern also served to camouflage the insect, with the holes creating a dizzyingly complex green array. I performed a few feeding trials by placing beetles in jars with different species of foliage, and found they only ate stinger leaves. In the absence of their preferred food, the beetles quickly died, confirming they were host specific.

With observations by volunteers, as well as my own climbing data, it became obvious that herbivores preferred young leaves. We saved a lot of field time by prioritizing our herbivore searches on young leaf surfaces, since insects rarely ate older ones. I also wanted to gauge how leaf toughness changed with age to better understand insect preferences. In a scientific journal, I read of a British plant biologist named Paul Feeny who had designed a gadget called the penetrometer for measuring toughness of oak leaves in Britain. It simulated an insect's jaws biting through leaf tissue and provided a measurement of how much pressure was required to puncture the leaf. I went back to Basil, in the university workshop, and together we built Feeny's gadget using the exact protocols. The penetrometer consisted of a beveled shaft that broke through a leaf surface with increasing pressure, created by the force of gently pouring water into a container. Young leaves broke with almost no pressure, but mature beech leaves were so tough that I had to place a three-inch-thick *Insects of Australia* textbook on the template and then add water to a large bucket. Earthwatch citizen scientists measured the toughness of dozens of species, both young and old leaves, confirming that young leaves were significantly softer than mature ones. Everyone chuckled at becoming a penetrometer expert, but the name predated me, so I was not going to change Dr. Feeny's terminology.

I decided my insect sampling had to include different seasons when new leaves were present (or not). But this required a few different types of collection methods. How does an arbornaut find tiny beetles in two-hundred-foot-high trees with vines twisting all over, hidden among masses of moss and lichens? First, I ordered a dozen rugged sweep nets, each with a wide mouth for catching the largest walking stick and a perfect-sized mesh that kept most critters inside while still allowing a collector to view its contents. I designed a repeatable field regime—ten full sweeps while walking (or gliding on the rope), to sample approximately ten cubic yards of foliage per sample. We netted bugs by day and night, in rain and sun, at high elevation and low, in sunlight and shade, and during leafing months and not. We found that nets were inadequate for sampling in the dark and best for sweeping through air or small branches. Harking back again to childhood nature camp, I recalled additional capture methods for insects: (a) nets to sweep through foliage and collect airborne critters such as flies, butterflies, moths, grasshoppers, beetles/weevils, and bees/wasps; (b) beating trays to shake branches and count what falls onto the sheet, including beetles, weevils, caterpillars, ants, walking sticks, spiders (not really insects but arachnids), and flies; (c) pooters to suck small creatures such as ants or mites into a vial using rubber hosing (these versatile sucking gadgets with a god-awful name were also a favorite necklace, adorning my field outfit in lieu of jewelry); (d) tanglefoot to slather on leaf surfaces, so its sticky goop captured alighting insects like boots in quicksand; and (e) light traps to attract nocturnal insects such as moths, beetles, flies, scorpions, and even spiders that come to eat bugs attracted to the light.

All these methods were ideal for citizen scientists to count moths, photograph weevils, shake insects from foliage onto beating trays, and conduct sweeps low on the forest floor and high in the walkway. My citizen scientists were especially thrilled, knowing they were contributing new information to science. Few ecological surveys of insect populations had been conducted in Australian rain forests, and certainly none in the canopy, so these efforts in the 1980s were a global first. The pioneering ecologist Dan Janzen from the University of Pennsylvania studied insect diversity in tropical forests, but his work was restricted to Costa Rica and mostly at ground level. Similarly, the innovative work of the Smithsonian Institution's Terry Erwin, who claimed millions of insects lived above our heads, fogged only one tree species in Panama, limiting its application to Australia. My Earthwatch teams were the first to conduct insect biodiversity surveys for a whole forest—not just in Australia, but anywhere.

Although leaves were the main focus of my thesis research, I grew pretty excited seeing insects crawling, munching, and flitting about the upper branches and marveled at the amount they consumed. To get a small glimpse into their spatial and temporal variation, I selected two trapping methods for comparing insects between seasons and forests: sweep nets by day and light traps by night. We deployed one hundred net sweeps and ten hours of light trapping as repeated units of sampling. Unfortunately, no single method existed to compare samples during both day and night because light traps weren't effective during daylight and sweep nets weren't safe in the dark.

Using light traps hoisted into the upper branches, our nocturnal catches averaged over two hundred collections per night. I will never forget one collection in the Antarctic beech stands, when the box holding the light overflowed with thousands of bogong moths—very nondescript brown, fat-bodied moths that were easily distinguishable from other colorful counterparts, but way too many to count! Such fieldwork provided a glimpse into the diverse array of colors, sizes, shapes, feelers, wings, textures, hairs, and classifications of arthropods. One trend emerged: in these evergreen rain forests, insect populations were correlated with pulses of new leaves. Cool temperate forest had one major, rapid leaf flush of Antarctic beech with a parallel surge in insects. Subtropical forest had multiple peaks of insects, paralleling its year-round leafing phenologies.

Insect collecting was essentially a numbers game, producing lots of data. It was never possible to identify everything in the field, but citizen scientists usually learned to recognize the specific orders of herbivores: Coleoptera (beetles and weevils), Lepidoptera (butterfly and moth larvae), Orthoptera (grasshoppers), and Phasmida (walking sticks), plus a few oddballs such as some Diptera larvae and a few Hymenoptera (for example, gall wasps). Through simple field classification, plus a few taxonomic keys and microscopes, we tallied baseline data of insect abundance and the proportion of herbivores. The results ultimately showed that insect abundance was greatest in the highest-diversity forests and in the upper canopy, not the understory. But herbivore numbers were greatest in the cool temperate rain forests, probably because this single-species canopy composition provided a reliable, all-you-can-eat salad bar of one greenery.

So, what does all this consumption mean for a tree's health? In general, the loss of leaf surface area can lead to decreases in wood production, photosynthesis, and reproductive capacity. In some cases, however, plant physiologists have shown that moderate herbivory can stimulate photosynthesis of the plant, like the way mowing a lawn can encourage renewed growth of grass. Some grasshoppers secrete growth-promoting substances into grass stalks when feeding, thereby ensuring their future food supply. Maybe other defoliators have a similar action to encourage the survival of their food plants? Unfortunately for us forest scientists, most research on plant-insect interactions has been conducted on annual plants in laboratories, making it difficult to extrapolate the results to long-lived trees in a forest. After delving into insect diversity and leaf toughness, I found myself thinking about feeding efficiency as an ecosystem concept. Some insects ate patches of leaf, causing the rest to become brown and die, whereas others fed very efficiently, so no leaf tissue ever turned brown and was ultimately wasted. It reminded me of humans—some leave uneaten food on their plate that goes to waste, and others do not. I wondered if the way insects feed—entire holes versus quilted bites that result in dead, wasted tissue—might impact the plant. So I designed a small experiment to examine two questions: Does the way an insect consumes a leaf affect the plant's ability to recover? And does moderate defoliation stimulate growth, as documented for some annual plants like grass? Since I could not conduct this type of experiment on full-grown trees due to the impossibility of weighing the entire plant's biomass, I resorted to seedlings in a lab.

I transplanted 130 coachwood seedlings from the forest floor into a controlled environmental cabinet back in the botany laboratory. They were all six months old, germinated in a light gap after a treefall. Each seedling was potted with similar soil, given identical nutrient treatments, water, and light regimes, and grew four months inside the chamber. Five were harvested to calculate average seedling size at the beginning of the experiment. The remaining 125 were divided into five groups: (1) controls, (2) one of every fourth leaf removed, (3) 25 percent of each leaf removed, (4) every second leaf removed, and (5) 50 percent of each leaf removed. In this case, I wanted to find out if different amounts of herbivory (none, 25 percent, or 50 percent) and chewing strategy (whole versus partial leaf consumption) affected growth.

After eight weeks, the 25 percent defoliation treatments showed the greatest growth, indicating moderate consumption appeared to encourage growth in coachwood seedlings. In contrast, removal of half of the leaf area (50 percent) was too excessive for recovery, and most of those seedlings died. The removal of one entire leaf (out of four) caused seedlings to reduce their level of regrowth, whereas 25 percent clipping of each leaf stimulated growth, like mowing grass. I found that a moderate amount of defoliation, even as much as 25 percent, stimulated growth in seedlings, but I have no proof that a seedling behaves similar to an adult tree.

Walkways have the magical effect of making a researcher invisible to the surrounding biodiversity, probably because wildlife feels relatively safe in the canopy high above the forest floor. It is a special feeling. Once, my team was setting up light traps in a Queensland rain forest when a flock of brush turkeys landed in the upper branches for their evening roost. As is often the case when birds settle for the night, they defecated on all of us, plus a rain of insects from those upper branches became dislodged and also fell. This humorous episode made me curious about the behavior of nonflying insects in tall trees. Do herbivores that can't fly starve if they fall to the ground? Or can they navigate back to their aerial salad bar? Rain forests were too tall and complex to experimentally remove caterpillars from foliage and observe their response, so my marine colleagues suggested a simpler system to address this caterpillar fixation: low-lying vegetation on coral cays. The Argusia bush (*Argusia argentea*) was a common shrub on One Tree Island, Sydney University's research station on the Great Barrier Reef off the coast of Gladstone, Queensland. In exchange for serving as a diving safety buddy, I spent a month on the island and undertook a caterpillar experiment. The larvae of the moth *Utetheisa pulchelloides* (Lepidoptera: Arctiidae) were host specific, only eating the Argusia shrub that frequented islands throughout the Great Barrier Reef. (Technically, only butterflies and moths are called caterpillars before they metamorphose; "grubs" and "larvae" are more general terms, including for beetles.) I had already visited One Tree Island and calculated that larvae ate approximately three square centimeters

(1.2 inches) of foliage per day, representing 2 to 5 percent of total leaf growth by the shrub. When placed in cages with other food choices, the larvae died if Argusia leaves were not offered, confirming they were host specific. To test whether caterpillars could navigate back to their host plant if displaced by wind or bird activity, I placed twenty caterpillars in a black bag to block out light-directional senses, and then positioned them either north, south, east, or west of the shrubs. They showed the greatest ability to relocate the host plant when displaced on the west side of the shrubs, but required thirty minutes to travel three yards back to the bush. Obviously, such slow travel would likely put them at the mercy of hungry birds and suggests they were probably not navigating but just randomly bumping into the bush. Similar to extrapolating a seedling experiment in the growth chamber back to the forest ecosystem, it is also impossible to link findings using shrubs to taller, more complex forest trees. Field biologists often ask questions at small scales, and then we remain limited to mere speculation when applying the results to a larger context. As a graduate student, many of the questions I posed offered a means to learn more about experimental design but not always to achieve significant conclusions.

On One Tree Island, I spent many hours in lively discussion, comparing herbivores in canopies to algal-feeding fish on coral reefs. My marine colleagues observed that certain groups of reef fish, collectively called feeding guilds, occupied the same heights in the water column or fed on similar types of algae. In parallel, certain weevils and beetles fed at the same height and specific age of leaf tissue. As scientists discover more about rain forests and coral reefs, we find overarching concepts—extremely high species diversity, extraordinary complexity within a three-dimensional grid, guilds of species, and coexistence of host-specific and generalist feeders. With encroaching climatic extremes due to human activities, interactions in both systems are increasingly at risk. Can species adapt fast enough to keep pace with changes in temperature or rainfall in forests and/or warming trends and shifting pH in oceans? Can organisms survive the rapid habitat degradation that many systems now face? Scientists work tirelessly in hopes of finding solutions to human-induced alterations threatening to collapse our natural systems.

My research documented seasonal canopy damage whereby insects decimated new leaves, and also confirmed the resilience of trees to higher levels of herbivory than previously estimated at ground level. Even though I calculated exactly how much defoliation insects caused, no field biologists have yet developed reliable techniques to accurately estimate the numbers of insect culprits doing all the chewing. I keep hoping some clever engineer might someday invent a detector to count every living arthropod in one cubic yard of foliage, but no luck. At a global scale, ground-based studies indicate spiraling declines in insect populations, with upward of 40 percent predicted to disappear in the next few decades. In one twenty-seven-year-long study using malaise traps (where insects innocently fly into a large cubic space surrounded by fabric, becoming captives) in Germany, flying insects decreased by 76 percent, with similar losses for Puerto Rican forests. Pollinator numbers are also in steep decline, yet approximately 85 percent of the world's flowering plants depend on them. According to the Food and Agriculture Organization of the United Nations, three out of four crops consumed by humans depend all or in part on pollinators. As a stark example of population decline, America lost one-third of its honeybee colonies during 2017 alone. A new and more urgent insectageddon threatens entire arthropod populations, with human actions like burning, agriculture, pesticides, and climate change driving many species toward extinction. Because scientists have not documented the ecology of insects as well as for larger animals, we don't really know the extent of their declines, and since insects are not as charismatic as dolphins or primates, there is little public outcry. This provides even greater impetus to improve our toolkits to study insects and their salad bar in the eighth continent, sooner rather than later.

⫸ Giant Stinging Tree ⫷
(*Dendrocnide excelsa*)

MY FIRST EXPOSURE to this gorgeous, fuzzy, and soft-looking leaf came during a solo exploration of Australia's subtropical jungles. When I first touched its brilliant emerald foliage, I had the sense of either grabbing a hot coal or having thirty wasps sting all at once. Toxic hairs lodged under my skin, and the pain lasted for days. I was even more surprised to experience the same excruciating burn when I picked up a dead leaf on the forest floor. Not fun! Even a one-hundred-year-old herbarium specimen can gave a noxious sting to an innocent botanist who handles it without gloves. One fellow student claimed touching the even more toxic Queensland Gympie-Gympie was "like burning yourself with hot acid and getting electrocuted at the same time."

Not much information exists about the ecology of these noxious giants, probably for the logical reason that they repel humans (as well as most other animals) from getting too close. In Australia, the family Urticaceae has one genus and seven species. Native to the Asian tropics where Urticaceae includes 2,625 species among 53 genera, many species feature toxic leaves and stems. An early Australian chemist named J. M. Petrie analyzed the chemicals of the Urticaceae in 1906 and confirmed that stinging trees were thirty-nine times more toxic than the small, shrubby nettles growing in temperate pastures of Europe and North America. Numerous chemical and physical stinging hairs cover the blades and petiolar surfaces, suggesting the evolution of morphological plant defenses against mammalian predators. Using scanning electron microscopy, I counted hair density in different ages of foliage and found highest density in young leaves before they expand. So I became extra cautious to avoid touching young leaves. Because the Urticaceae family predominates in the Asian tropics, toxic hairs may have evolved to protect trees from monkeys and other mammal predators, although they do not prove such a good defense against a few resilient insects.

Australia's giant stinger, *D. excelsa*, with its status as an emergent, exhibited leafing for eleven months of the year, but over 60 percent of new leaves expanding during summer (January to March). This provided a reliable feast for one host-specific insect, called the stinging tree beetle (*Hoplostines viridipenis*), which developed a unique ability to digest the toxins and navigate its spiny surfaces. A toxic surface also gave the beetle protection from bird predation, so it fed brazenly on the upper leaf during daylight. No fear! Because stingers grow fast and tall as opportunists in light gaps, they are relatively short-lived with soft wood, reflecting a minimal investment in longevity or toughness but a strategy of fast growth to compete for a canopy spot. Of my five research species, stingers had the shortest lifespan, only seven months, and each branch put out a generous flush of 8.3 leaves annually. During that brief lifespan, beetles ate approximately 32 percent per leaf, and when calculated over a twelve-month time frame, totaled 42 percent annual crown loss. Holy cow! Leaf fall occurred all year long, and leaves decayed within four months, the fastest rate of any species measured. I saw that an underlying strategy of giant stingers was to grow fast, grow cheap, get eaten, and turn over quickly. When a light gap opened on the forest floor after treefall, their seeds, which were abundant in the soil, germinated quickly, continuing their strategy to achieve a prominent yet short-lived spot at the top. In this fashion, they grew fast but ultimately lost out to slower-growing species whose wood was stronger and more long-lasting.

In addition to a resilient, host-specific beetle, giant stinging trees housed gazillions of sap-sucking aphids (*Sensoriaphis furcifera*). At times, up to one hundred aphids per leaf created almost wall-to-wall bug sucking events, after which the remaining tissue dried up and died! On rare occasions, also observed was munching by the spur-legged phasmid (*Didymuria violescens*), a relatively large walking stick whose digestive system was tough enough to tolerate the toxins. Flying foxes sometimes roost in the branches, but relatively little other biodiversity. Some animals managed to eat the fruits, including an occasional possum, bird, snail, frog, or lizard.

Two notorious northern relatives of the giant stingers are the shiny-leaf stinging tree (*D. photinophylla*) and the Gympie-Gympie bush (*D. moroides*), named by nineteenth-century gold miners working in the region of Gympie, Queensland. The latter was even more toxic than *D. excelsa*, and a shimmering green-black beetle (*Prasyptera mastersi*) exclusively ate this species in a lacey feeding pattern similar to the giant stinging tree beetle. Red-legged padymelons (*Thylogale stigmatica*) also ate that fiery foliage, sometimes devouring an entire bush, which defies explanation. This close relative of kangaroos and wallabies somehow digested the toxic greenery, and no other animals competed for its diet. Out of thousands of rain forest plants, it seemed extraordinary that so few species evolved stinging hairs, suggesting that factors other than physical defenses may be more important in a tree's competitive world.

5

DIEBACK IN THE OUTBACK
꙰꙰

Juggling Marriage and Investigations of Gum Tree Death
in Australia's Sheep Country

THROUGHOUT MY DOCTORAL YEARS studying rain forests, I spent a lot of time (best guess ~1,500 hours) driving along old logging roads and stopping the car every 500 feet or so to sample leaves. This was a great way to spot-check for insect outbreaks or observe unusual foliage anomalies. Occasional sampling along roadsides was a shortcut to avoid rigging tall trees. These edge leaves were bathed in sunlight, physiologically similar to the sun leaves high above, but I could collect them by simply jumping out of the car with a plastic bag and randomly clipping thirty specimens. I often traveled with my trusty friend Hugh, who drove our borrowed University of Sydney Holden station wagon so I could hop in and out easily. As graduate students who both encountered hazards in our respective work (intertidal shores and treetops), Hugh and I were safety buddies for much of our fieldwork. In field biology, this is essential. Risking my life with tidal surges on rocky shorelines, I counted barnacles in between crashing waves for Hugh, who studied their population dynamics. In exchange, he served as a climbing grip, also known as "dirt" (because he stood on the ground to monitor for safety), especially important for night climbs.

The Sydney University car fleet consisted entirely of Holden station wagons, big enough to hold all our field gear. In late twentieth-century Australia, two lifestyle commodities were symbols of economic success: a clothesline called the Hills Hoist (a truly ugly metal contraption erected in backyards to hang laundry) and a Holden station wagon (rugged rectangle of solid metal that could hold countless kids and heaps of food and "tinnies," or cans of beer). These rugged vehicles were ideal for long distances involving plenty of dust, frequent cracked windshields, and enormous vigilance for 'roos. In between navigating all those driving hazards, my awestruck eyes took in the gray-green expanses of gum canopies, stunning blue mountains, endless empty beaches, crimson flashes of flowering flame trees, and occasional roadside sculpture. Aussies have a great sense of humor, and their roadways are dotted with enormous monuments, collectively called Big Things and dedicated to charismatic organisms. Hugh and I kept a checklist during our long drives to and from research sites, hoping to spot all 350 Big Things across the country: big banana, big beer can, big merino, big prawn, fat strawberry, giant dog, big pineapple, and oversized lobster. A town in Queensland tried to erect a giant bust of the much-despised invasive cane toad in hopes tourists would want to come photograph such an infamous creature. Not surprisingly, the local council voted it down.

One fateful day, as we drove slowly along a logging track, another car raced around the corner, smashing into our front grille, and I crashed against the dashboard. Seeing my nose bleeding and face disfigured, three district engineers leaped out of the fancy government SUV that had rammed our university station wagon. In no time, they delivered me to a regional hospital in Armidale, New South Wales, where I was later dismissed with multiple facial stitches and bandages. Where to go while I recuperated until the hospital could remove my stitches? In this rural area, I remotely knew the name of one biologist, a distinguished professor at the local university. Professor Hal Heatwole was renowned for his work on ants, sea snakes, and Australian ecosystems.

Although we'd never met, I'd followed his publications carefully because he was an unusual breed of scientist who tackled multiple disciplines. I looked for his name in the local phone book and made a cold call. Soon after, I was propped up with pillows on Hal's living room couch. We became fast friends, and during my weeklong recovery, plotted out a new research plan. Hal was excited about my whole-tree perspective and wanted to apply it to the regional incidence of eucalypt dieback. In rural New South Wales, a protracted decline in health and vigor of many gums plagued the outback landscape. Its aftermath was harsh, with cattle and sheep huddled beneath skeletons of former trees that had lost their leaves and offered little shade. Most researchers to date had either sampled for fungal attacks on soil or measured the water table for salinity, but no one had searched the canopies for clues. I brought a new perspective to forest health, and we both agreed it would be a good use of our collective expertise to help rural landowners solve their serious tree Armageddon.

Having just submitted my PhD, I was proud to have proved the department chair wrong after he, only three years before, had suggested I would just get married and have kids. I was not required back at Sydney until six months later, to pick up my bound thesis and attend the graduation ceremony. But my thoughts were already focused on "What next?" and the gum tree dieback posed a significant ecological problem to which I could apply my canopy-access skill set. It took one day by bus to reach the University of New England in Armidale, New South Wales, Australia's only rurally situated institution of higher learning, where Hal taught. From the bus window, I gazed at miles of dry pastures dotted with thousands of sheep and dead tree skeletons. Gum trees were a prominent feature of the rural landscape. Farmers needed pasture trees to shade their livestock and prevent soil erosion. The timber industry also relied on eucalypts for sustainable harvest to provide livelihoods. Even the Australian government loved gums: they were iconic trees and created landscapes that encouraged tourism, a growing part of the rural economy. Native birds and insects needed their crowns as homes. And another iconic Australian resident, koalas, lived by eating gum leaves, to the extent that some locals thought koalas were killing the trees by overeating foliage. This dreadful rumor even incited some graziers to start shooting koalas on sight. So not only did conservationists and farmers have pressing reasons to solve the dieback, but koalas did too!

This same region of Australia had experienced a large-scale tree death in 1886, an episode recorded in a few rural diaries by farmers at the time. As a result of that single event, intervals of tree dieback were calculated as one hundred years apart, although no ecologists existed on the scene to assess the trees back in the late 1800s. When the dieback resurged in the early 1980s, ecologists in Australia (as well as around the world) took careful note of the demise but did not apply the term "climate change," which is now recognized as the ultimate underpinning of emerging death syndromes of many forests, including Australian gums. A few decades later, scientists understand that some countries subject to climate anomalies around the Pacific Rim like Australia act as early warning systems of global climate change as it accelerates in frequency and extremes.

In outback Australia, patterns of rainfall had always been the make-or-break factor for life on the land. Like scientists, ranchers did not use the term "climate change" in the 1980s, although the rural sector was painfully aware of increasing droughts and hot spells, plus the related scourges of more frequent fires and a shortage of grass for livestock. During these accelerated dry spells, insect populations exploded, consuming vast quantities of leaves that trees need to thrive. Temporary recovery of dying individual trees sometimes occurred, with small branches (called epicormic shoots) leafing out up and down the trunks as a last-gasp effort of small sprouts, or tiny stump sprouts attempting to regrow at the base of otherwise skeletonized trees. Fieldwork on tree dieback in the outback during the 1980s was the precursor episode of extreme climate change that has taken center stage in Australia some thirty years later.

"Gum" is the common name for many species of the genus *Eucalyptus* native to Australia. Approximately 555 species exist in this genus, but the number changes if taxonomists decide to lump together or split apart species due to DNA analyses or strong opinions about physical characteristics. The battle between lumpers and splitters continues to plague the world of taxonomy. Not all of the 555 or so species of *Eucalyptus* were subject to this sudden death syndrome, and different causes were identified in different regions. In Western Australia, the death of jarrah (*Eucalyptus marginata*) was clearly linked to a root fungus accidentally introduced from Indonesia. Increased soil salinity was implicated as a major factor in South Australian episodes, killing several gum species. But in New South Wales, no obvious factor was implicated. Rather, it seemed to be an insidious death with many suspected causes: fungal disease, insect defoliation, koalas eating gum leaves, drought, nutrient imbalance in the soils, application of fertilizers (especially superphosphates aerially sprayed by farmers to stimulate grass growth), depletion of water tables, increased salinity, clearing, reduced crown cover, and overstocking of grazing animals that ate seedlings and ring-barked the trunks. As is often the case with disease, a complex of potential causes makes it tough to tease apart any one factor from the rest.

Eucalypts have many extraordinary characteristics, but their survival is a true lottery. During a good year, one mature tree might produce five million seeds. The seeds are light and wind-dispersed, and many require fire as a natural method to split open their woody encasements, called gum nuts. Seeds fortunate enough to be blown into a place with moisture, soil, and access to sunlight will germinate, but less than 1 percent have such luck. After that, multiple hazards threaten a seedling's childhood—drought, trampling by hooves, extreme winds, excessive sunlight, fire, consumption by insects or livestock, flooding, and human clearing practices. If a seed germinates in a favorable location, it grows into a sapling within a few years' time. Some species build up fire resistance by producing layers of peeling bark, and others develop a unique taproot called a lignotuber that can access groundwater far below the soil surface as well as resprout after fire. With the onset of a continent-wide dieback in the 1980s, very few (if any) gums were observed growing, either as adults from seed germination, or as sprouts around dead trunks. In short, the landscape was becoming increasingly barren.

Once I was set up in Hal's laboratory at the University of New England, I wrote a grant requesting funds as a postdoctoral researcher, which is a fancy term for a stepping-stone position after obtaining a PhD. I rented a small apartment near the university and found a roommate to share costs. Judy and I met at a short course on massage, where we both obtained a masseuse diploma so we could work to pay a few bills. I obtained the grant in three months; it provided three years of salary and field equipment to climb eucalypts, so the massage career was short-lived, though it introduced me to a circle of loyal women friends. Using my successful methodology from rain forests to survey the whole tree, I hypothesized that insect defoliators represented a "last straw" in dry forest canopies after an onslaught of prior environmental degradation from other factors. The farmers claimed it had been warmer and drier during the 1980s, and that they were applying increasing amounts of fertilizer to the pastures, so insects might be attacking trees that were already stressed. I needed to locate some study trees to test the hypothesis. How could I persuade a few ranchers to allow a fair dinkum scientist to climb on their properties? The easy solution, according to Hal, was to visit the local pubs with a slingshot and tell a few good tree-rigging yarns. I knew from my time in the rain forest regions that when the pub lights go on, rural farmers flock there just like moths swarming to light. Many rural watering holes still had women-only sections, so I had to enlist a few male colleagues to accompany me to the men's side. Hal and a few of his students were happy to oblige. Lo and behold, the pub strategy was successful. In no time, I had a few stations (Australian term for ranches) on which to climb and study canopies.

One property was owned by a young grazier named Andrew. My rural girlfriends joked he was probably the only eligible bachelor within a thousand square miles! Andrew was a good-looking, funny, and creative farmer with a love for nature. He blamed beer for sidetracking an academic career, but most sheep and cattle ranchers rely for their success on a network from the local pub, not from a dusty college diploma hanging on the woolshed wall. Our courtship mostly amounted to my assistance opening gates when Andrew moved sheep around his five-thousand-acre property, bringing lunch to the woolshed during shearing, and an occasional pub dinner. But we talked for hours about sheep, landscapes, trees, and life in the bush. I was thirty years old and had been focused primarily on field research for twelve years, but always wanted to have a family. I could hear the ticking of the biological clock in between the crunching of insect jaws on my beloved leaves. And after our first meeting, his parents dropped hints about both of us becoming too old to provide them with grandkids! Andrew and I thought we could turn his family farm into a model property, with my expertise to restore native trees and his sense that a healthy landscape would ultimately produce healthier livestock. We fell in love and got engaged, so he called my parents in upstate New York with the news. Oops—he overlooked the time zones . . . it was 4:00 a.m. . . . off to a bad start! Mom cried. Dad, always a gentleman, acted polite. We were married in Australia and had an after-party in our woolshed. None of my family attended. Mom shed more tears, and I was disconsolate to live ten thousand miles away. But it seemed an ideal world—love, marriage, imminent children, and thousands of dying trees.

At the ripe old age of thirty-one and immersed in my first career assignment, I moved onto Andrew's family sheep and cattle station, one hour from the university and a huge distance from any biological family. Our nearest town, Walcha, an aboriginal term meaning "watering hole," boasted over 200 times more livestock than people. A billboard outside town announced, WALCHA—SHEEP 760,000; CATTLE 120,000; PEOPLE 4,000. This region was also the epicenter of dieback. The sign probably should have included DEAD TREES 500,000. The main street housed four pubs, one small grocery store, a post office, a pharmacy, three banks, and three rural supply stores. Banks were important for borrowing against droughts and fluctuations of agricultural markets, and pubs were critical for drowning sorrows or celebrating weddings and births. Australian graziers were a stoic lot, proud descendants of convicts sent via ship from Britain, and trained to mix their blood with soil. They frequently dedicated eighteen-hour days to livestock, fences, watering holes, and pastures. During their little time off, farmers loved to drive around the local district as sticky beaks, spying over fences to find out what their neighbors might be doing better. When a grazier needed a mid-morning smoko, their wives were ready and waiting in the kitchen. The "Sheilas" stayed home to cook lamb chops, swat blowflies in the kitchen, keep the toddlers safe from venomous snakes, and generally manage the home front. It was not an easy life, and the rural culture frowned upon women who fancied themselves intellectual. It was not long before my in-laws labeled me as a bluestocking, that derogatory term for any girl who embraced academic pursuits.

But my farmer husband proudly married me despite, and maybe also because of, his wife's tree-love. I was certainly the only qualified arbornaut in all of Australia, and our rural landscape suffered from a botanical malaise of enormous economic, ecological, and emotional magnitude. I was determined to figure it out. The ghost-gray skeletons of hundreds of dying gums, visible from our kitchen window, haunted me every day. Australian farmers needed trees to shade their livestock, conserve soil, house native birds and insects, and keep the natural systems humming. But none knew what caused the deaths or how they could reverse the serious degradation of their properties. These questions gripped me both as a scientist and a newly minted farmer's wife.

Whereas many college classmates married into wealth, beach houses, or big yachts, I married into five thousand acres of dying eucalypts and felt blessed to have this incredible treasure. I called it our "tree dowry," because I got thousands of trees via marriage and, in return, my husband got knowledge about the trees. Our five thousand-acre sheep station, very sparsely dotted with trees, supported as many as fifteen thousand head of fine-wool merino sheep, but only if it rained and the grass grew. Like most graziers in outback Australia, the livestock and the fencing were more important than the homestead. So we always bought a new gate or a roof for a shed before a dishwasher or couch. Our newlywed home was actually a rundown cottage at the edge of the family station; it had been uninhabited for a decade. The luminescent orange kitchen was tamed by my white paintbrush, but I could never get the window screens tight enough to keep out hundreds of sheep blowflies that invaded whenever lamb was cooking. A few years later, I was horrified to find blowfly grubs crawling all over our newborn son during naptime; larvae had infested the wool crib blankets. There was not much I could do except pick them off, in tears.

Our cottage was essentially a large, isolated wooden shed designed to blend in with a certain bush beauty. The wooden floors creaked, plastic strips fluttered from the doorjambs in an unsuccessful attempt to keep out flies, and scary-looking ancestral portraits hung in what the kids affectionately called our "hall of horror." In those years, graziers' wives were proud of their shiny laminated kitchen floors, plastic kitchen chairs, threadbare upholstered couches, and instant coffee. Yet despite such humble trappings, an extraordinary warmth stitched all the rural homesteads together across great expanses of thousands of windblown acres. My new husband was part of a unique heritage, one that had passed down to him amazing abilities to read the landscape. I admired how he and his father tamed the bush, opened and closed thousands of gates to move livestock between paddocks, ingested their fair share of dust and blowflies, repaired innumerable miles of fences, maintained our sixteen miles of telephone lines, trained a small army of sheepdogs, and endured the roller-coaster rides of wool, lamb, and beef markets. Despite a quixotic love of trees and a bluestocking nature, I quickly amassed a cadre of wonderful friends, all wives at neighboring stations. We met often for coffee, for playdates when our babies were born, and thought nothing of traveling forty or fifty miles to visit one another for a few hours to share our trials and tribulations. They admired my worldliness and intellect, and I loved their humor about rural life and practical solutions to outback housewifery. Thirty years later, I still keep in contact with my sisterhood of Australian women friends, most of whom still live on the very same sheep properties. As a graduate student at Sydney University, I had taught night school to blue-collar workers for three years, until I no longer felt like throwing up when giving a lecture. I credit night-school teaching for bolstering my courage to speak in front of others. Throughout my career, the ability to communicate with locals, and build trust, became an asset far more useful than a wheelbarrow full of technical publications. Speaking plain language was key to building relationships with the farmers whose trees were at risk.

Life on the farm was tough, but I loved awakening at dawn to the demonic call of the kookaburra, the lyrical songs of magpies, and the bleating of several thousand sheep. I worked tirelessly to cook wonderful meals, shopped during lunch breaks on days when I went to the university, and often raced home from the office, exceeding the speed limit, to ensure my farmer husband had a hot dinner awaiting. I was fortunate to never crash into a kangaroo on our rural dirt roads given the rush to switch gears from academic to wife. I wanted to be the perfect spouse in every way, but my in-laws were extremely critical about their daughter-in-law's bluestocking behavior. It was a joke among the neighborhood girls that my mother-in-law was happy to babysit if I had a hair appointment but not if I went to the university library. So I tried to conduct field research in the most unobtrusive fashion. In my heart, I believe she was trying to mold me into a good farmer's wife, but it was impossible to shed a life of tree-love. And I had already launched the eucalypt research well before marriage so felt determined to follow through and solve this rural epidemic.

From initial observations of rural eucalypts, insect outbreaks appeared to represent a last battle for a tree, after multiple skirmishes with drought, heat, and human-induced stresses. I knew from prior research that moderate levels of about 25 percent annual herbivory did not kill rain forest trees, so if insects contributed significantly to mortality in gum canopies, then the pasture trees had to suffer much higher defoliation. My research methods in gum trees would be similar to the rain forest: determine multiple species to study, find branches at vertical intervals up replicate crowns, and label leaves to monitor herbivory and mortality over time. But climbing in dry forest trees was a whole lot different from climbing in wet forests. For one, the landscape was more open and desolate, with winds and harsh sunlight beating down all day. It was best to climb at dawn or dusk, to avoid sunburn and a gazillion blowflies swarming around my sweaty face during the sultry daylight hours. As compared to rain forests, the limbs of fast-growing, sunbaked eucalypts were extremely brittle, so selecting a safe branch to support the climbing rope was critical. My rule of thumb was to rig a branch greater in circumference than the climber's thigh. This dimension was probably two or three times larger than required, but it gave me peace of mind. Angry bulls roamed in some pastures, so I had to keep an eye on livestock as well as branches. One unexpected joy was the occasional koala (*Phascolarctos cinereus*) feeding in one of the study-tree canopies. Incorrectly called koala bears (they are not even related to bears), these arboreal marsupials exclusively eat eucalypt foliage, but only from a handful of species. Over many years of climbing, I was thrilled to occasionally share tree crowns with slow-moving koalas after they had consumed a big dose of gum leaves, toxic to most animals due to their essence of volatile eucalypt oils, but not to the koalas, which had adapted to digest this chemical salad. They could not fly away, nor did they slither or jump or scoot down the trunk; instead, they usually just looked at me as if I were an alien. But they could really sprint on the ground when switching trees, and several times I watched them outrun our sheepdogs.

After four years of rain forest climbing, I rigged eucalypts quickly and expertly. I chose five species native to northwestern New South Wales: New England peppermint (*Eucalyptus nova-anglica*), Blakely's red gum (*E. blakelyi*), New England stringybark (*E. caliginosa*), black sallee (*E. stellulata*), and mountain gum (*E. dalrympleana*). These species were all observed by farmers as victims of the dieback syndrome, and as isolated remnants on farms, their crowns were important habitat for biodiversity and essential for livestock shade. In other words, farmers really cared about their fate. It made sense for me to work in close collaboration with the graziers, creating solutions relevant to the health and ecology of their landscape. They came to respect me because I shared their love of the land, and this was invaluable to developing trust. Throughout my career as an ecologist, I have observed colleagues who stayed aloof of the locals and subsequently never bonded with communities where they worked. This can result in conflict and definitely leads to public distrust of science. I had to overcome the prejudices that farmers rightfully felt toward scientists, as well as prove it's okay for women to climb trees.

After marrying Andrew, I continued canopy work on rain forests, but only part-time. Only a handful of Australian arboreal scientists existed, so it was important for me to occasionally join expeditions to the wet tropics. During the spring of 1984, I found myself in North Queensland surveying the biodiversity of a tropical forest, and not surprisingly as the only female. On day three, I became seriously dizzy after climbing one of the forest towers to collect fruits, which was unusual after many years in the canopy. My body felt different. After work, I walked to the local drugstore near our remote motel and found a Penguin book called *On Pregnancy*. I discreetly brought it back to the motel and read for most of the night. The symptoms described in the book matched mine. I was secretly excited but had no mobile phone to call Andrew and speculate. Upon arriving back at the farm ten days later, I made an appointment with our rural GP, who gave me a pregnancy test, which proved positive.

As a pregnant scientist juggling work and farm life, the biggest hurdle turned out to be my in-laws. In their eyes, having children was a wife's primary role. When I became pregnant, it brought great joy to them as well as to Andrew and me. I painted the nursery, sewed pillows and quilts, and did everything a good housewife should do—even as I continued to quietly measure trees, set up litter traps to catch leaf litter and insect frass, review scientific literature, and continue postdoctoral duties at the University of New England. I desperately wanted to keep up a profession, not just because I had trained so hard to become a doctor of botany, but also because it provided a little personal income for things like buying a ticket to visit family or attending a science conference. The farm was financially successful, but most of the proceeds went into fencing and livestock vaccinations, certainly not to the daughter-in-law's research or her personal travel inclinations.

Because of my expanding tummy, it became awkward, then impossible (and downright dangerous) to climb with ropes and a harness. I needed a creative solution other than my simplistic, original single-rope techniques involving a slingshot and harness to carry out canopy fieldwork for the better part of nine months. So I borrowed a new piece of equipment from the agriculture department, a cherry picker, which allowed me to ascend with ease into the eucalypt treetops. As compared to rain forests, these dry stands were about half the height and trees were more widely dispersed, allowing fairly easy access. This slick gadget required no prolonged slingshot efforts, extraordinarily little perspiration, and the ability to "drive" up and down in the bucket by operating a few gears. It was truly fun! It would never have worked in a wet rain forest, however, without dry soil and open ground to navigate the trailer under each tree. Throughout those nine months, I devoted many waking hours to finishing up year three of foliage measurements, totaling 5,623 leaves (of which 2,543 had flushed after year one). On average, my five species suffered from 23 percent to 61 percent annual leaf consumption. But some of the marked leaves were entirely eaten by scarab beetles (*Anoplognathus hirsuta*), also called Christmas beetles because they were abundant during the Australian summer month of December. These voracious herbivores not only consumed entire crowns of New England peppermint trees, but during one season they ate three subsequent flushes, technically 300 percent leaf surface area per year! This repeat attack and extraordinarily high herbivory ultimately killed the peppermints as well as several other local species during the third year of sampling, proving that insects were a last stressor leading to death.

In my obsession with how insects impacted the aerial salad bar, I almost overlooked what they might be doing to the rest of the tree, especially the root systems, until I noticed Christmas beetles living in the soil during their grub phase. It seemed important to document their underground activity, to essentially return to that big toe of the trees and determine what was happening below. But how does a field biologist measure root damage? As it turned out, there was only method: demolish the tree and examine its roots. For this activity, as with earlier canopy research, I recruited Earthwatch volunteers, whose eyes and ears made the work more comprehensive. They stayed in our rustic shearers' huts and helped map both above- and belowground insect damage to gums. Our farm equipment was a boon for root research because a tractor could effortlessly excavate soil to expose roots and beetle grubs. First, using a cherry picker, the team carefully harvested the aboveground parts of one healthy New England peppermint and one identical-sized dying individual approximately one hundred feet away. Every cubic yard of foliage, branches, and trunks was mapped, cut, and bagged or labeled for analysis in our woolshed laboratory. Then we used the farm tractor to extract the roots. My farmer husband carefully dug trenches in and around the two trees so that the citizen scientists could use careful hand-sieving to extract all the roots. We all literally played in the dirt for several days. The results were astounding! Thanks to John Deere plus a team of citizen scientists, we calculated the magnitude of both canopy herbivory plus root damage by Christmas beetles. We weighed cookies (cut sections of trunk), counted every single leaf, and measured the defoliation of each tree. A healthy peppermint had a total canopy area of 161 cubic yards, representing just over 150,000 leaves. In contrast, the dieback tree had one-third of the leaf material, distributed across 90 cubic yards and totaling only 60,000 leaves. Borers had also attacked branches, eating away 19 percent as compared to a more moderate 5 percent loss in the healthy tree. Root damage from beetle larvae was even more vivid: the dying tree had only 20 percent remaining root matter as compared to its healthy counterpart. In short, Christmas beetles and their grubs were consuming peppermints at both ends. It was a double whammy—grubs fed underground and then emerged as adults to eat the leaves.

This research measuring insect damage to both roots and canopies was completed only because of my mom's heroic assistance. Midway through pregnancy, I developed dreadful vertigo and had to forgo all physical work, including cherry-picker activities. My intrepid mom came all the way over from upstate New York specifically to cook for the Earthwatch teams. In those days, it was at least five flights from Elmira to Walcha, with stopovers in Pittsburgh, Los Angeles, Hawaii, and Sydney to refuel and change planes. We laughed, because she outright refused to replace me in the cherry picker, and she certainly did not appreciate the swarms of blowflies in the kitchen, nor the venomous snakes in the grass. But thanks to her efforts, volunteers were well fed as they measured every inch of those eucalypts.

During pregnancy, I gained fifty pounds. It was considered a sign of good health in the outback for pregnant cows and ewes to gain lots of weight, so I guess the same rules applied to humans. Ten days past the due date, I was not only enormous but also anxious. For three afternoons in a row, I stumbled along the ruts and furrows of our back paddock behind the farm cottage. My mother-in-law insisted that this type of rough walking would incite labor. It was 1985 and farmers were all lamenting about climate extremes wreaking havoc on the agricultural and grazing industries. In a desperate effort to stay ahead of El Niño, we had just planted oats. Our sheep could not live through another season with the native grasses so sparse, so we hoped to offset their impending starvation with a higher-nutrition oats crop, but even that was a gamble because it would need to rain at least a few times for the oats to grow. I tripped and flailed through the rock-hard crusts of dry soil flung onto an undulating landscape by the plow, big enough blocks to break a leg if I was not careful. There were no artificial inducements for labor in our twenty-three-bed outback regional hospital, and despite the absence of technology, I had absolute trust in our country doctor. His simple practice offered no ultrasound, no election for an epidural or cesarean delivery—but many years of practical experience. If complications arose, I would be flown to Sydney via the Flying Doctor Service.

Eddie was born after thirty-six hours of labor. No drugs, no scans, no modern equipment. Most of our cattle did not undergo such an inhumane experience, as my girlfriends murmured. After the first twelve hours of labor, when it seemed obvious my progress was not accelerating, our faithful general practitioner wisely went to bed. I was relieved that at least one of us might get some sleep. At 4:00 a.m., the delivery table at the hospital collapsed when the nurse accidentally bumped it, and I tumbled to the floor. My handy husband got his toolkit out of the truck, fixed the broken joints, and lifted me back onto the repaired table. Some twenty hours later, after excruciating contractions and no painkillers, Dr. MacKinnon said I had his permission to swear. I was so exhausted my brain couldn't process anything, but I needed to muster one final burst of energy to deliver the baby so meekly said, "Jeepers creepers!" For months afterward, the entire community chuckled over that so-called expletive. After an overwhelming amount of breathing, pushing, and tapping an innermost fortitude, Eddie was born from a breech position, weighing almost nine pounds, but happy and healthy. I required a lot of quilting, as the doctor called it, from excessive tearing during childbirth. In the outback, one can only hope for good health and good luck because the technological trappings of modern medicine are many hundreds of miles away. I spent a week in our rural hospital, barely able to walk after all the stitches. But on the plus side, there was no rush to leave and no suspicion of baby substitution as occasionally reported in large urban facilities. In fact, no other babies were born in our rural hospital over the entire month, so Eddie was spoiled by all the nurses. When I first got home, some American childhood friends making a round-the-world trip showed up at the local gas station while driving from Sydney to Brisbane, hoping to visit. The attendant smiled and told them I never answered the phone between 10:00 and 11:00 a.m. because I was nursing the baby. Such was the level of close-knit knowledge shared in rural communities— everyone knew everything, or so it seemed. And I loved that sense of connection.

After the birth of our first son, Eddie's dad was proud and my in-laws were beaming, as was Eddie's Australian great-grandfather. He and I were remarkably close, sharing a love for birds, and despite his age of eighty-six, he drove every afternoon from his house a mile up the dirt road to have tea with me and watch birds at the kitchen window. He died several months after Eddie was born, and I secretly believed he was merely waiting for the arrival of a great-grandson before departing this earth. Even in the late twentieth century, male children were the major currency for successful inheritance of a working farm. Edward Arthur was named after both grandfathers, which were big shoes to inherit. Struggling with a newborn, I was hit hard by cultural attitudes confronting me on all sides, especially from the in-laws, who made it clear that a daughter-in-law's role was at home with the baby. Period. There was no childcare within a hundred square miles, and no immediate family offering to help. So I was pretty much locked in as a full-time "mum." A few colleagues at the university noticed my absence, essentially those researchers studying bird populations and tree-planting trials and pretty much comprised our entire dieback team.

When I asked my advisor if I could ever resume field research after having a baby, he simply suggested returning after one week off. Fat chance! (He also admitted he was part of a generation where his wife did all the child-rearing in their household.) Fortunately, I had a backlog of data that could be analyzed at home.

Housekeeping for a rural farm operation was already pretty much a full-time gig, but it became double duty with a baby. In the absence of a clothes dryer, the terrycloth diapers needed hanging on that backyard Hills Hoist. I cooked lamb a hundred different ways, and we ate our homegrown livestock almost every day. Baby food required cooking, straining, and sieving; it was not available in cute little jars at the grocery store. Men needed morning and afternoon teas when they dashed home from the paddocks, and because I made real coffee (my mother-in-law preferred instant), our kitchen was the favorite stop. Constant household battles were waged with blowflies, water shortages, dust, and venomous snakes. We found a family of red-bellied black snakes living in our outhouse next to the kitchen; I was not amused, but the men all laughed. We once found a dead sheep in our water supply and smaller bits of biodiversity (aka scum) lived in the roof gutters that drained into the belowground water tank. In hindsight, we probably ingested amazing microbes living in the outback, which may now confer immunity to some unknown pandemic. But there were also immediate dangers, such as when Eddie nearly grasped a brown snake around its neck while helping me turn on a hose in the garden. Both faucet and snake were brown and erect, side by side, but a brown-snake bite could kill such a tiny person. I grabbed Eddie in the nick of time, dashed indoors, and returned with the family shotgun. Andrew taught me how to shoot snakes, although I never really intended to do so. But this encounter made my blood boil. A venomous snake threatening a tiny boy in his own garden. The mom instinct is strong! Lucky for that snake it disappeared before I could take aim. At the best of times, I appreciate snakes, and have even dedicated research time to these important creatures. But as a young mom, I could not reconcile poisonous snakes and toddlers slithering in the same space.

When James was born sixteen months later, the nurse apologized with a wink for "accidentally" breaking my water with her fingernail. Remembering the long labor with Eddie, she could not bear to see me repeat such a prolonged episode. This time, after only ten hours of labor, James was born and immediately cried with hunger. He was big and strong and seemed ready to devour a steak. Having a second boy catapulted me to greatness in the eyes of my in-laws. Despite the scarlet s on my chest for SCIENCE, I briefly became a source of pride and joy. Birthing two boys was akin to winning an Oscar, ensuring succession of the ranch. Our sheep property was entering its sixth generation, all through male bloodlines. Andrew was the only male grandchild, so his life had been predetermined in many ways. Those hundred-plus years of family succession had seen their share of droughts, disease, invasive weeds, rabbits, snakebites, deaths, and financial struggles. Life on the land is truly tied to the weather, making it a lottery for survival and success. Whenever Andrew and I went to a pub, the graziers talked about weather, livestock, and little else; women mostly exchanged recipes and child-centric stories. A few pubs in our district excluded women completely, only offering a ladies' waiting room. When, on rare occasion, I accompanied my husband into the pub, the men looked at him even if they were talking to me. Such behavior reaffirmed an unspoken sense among men that women were inferior. One weekend, we went to a party hosted by one of Andrew's mates. Standing around a barbecue in the backyard, the host grabbed me in a very personal place. I immediately bolted back indoors. Unfortunately, the sliding door that was open when I came outside had since been closed, so I smashed into the glass, landing on the ground with my face bleeding. I ended up in the hospital getting stitches. When I whispered to Andrew about the incident, he felt it would not do any good to share the truth of the groping incident. Instead, everyone simply chalked it up to a woman's weakness with liquor. Although I was tempted to be angry, I had to accept this as a liability of my gender and the culture into which my husband was raised.

Not surprisingly, two toddlers turned my life upside down, especially in the absence of disposable diapers, day care, or convenience foods. Like their mom, the boys learned a great deal from trees in their childhood. We had no near neighbors, sidewalks, or playgrounds, so we spent lots of playtime observing nature around the sheep station. I taught them to sniff the scented juvenile leaves of New England peppermint, hide behind stringybark trees with their hanging bark strips when kangaroos approached, spot koalas feeding in eucalypt crowns, and watch a bowerbird collect blue fruits for his lovingly constructed courtship boudoir under one of our garden elms. What solace for me that my husband's grandfather had planted English elms along the driveway. Now their giant green canopies provided deep shade in an otherwise parched landscape and reminded me of the elm growing through my childhood lake cabin. Both boys learned to count using gum nuts, smell the fragrances of eucalypt oils, and fine-tune their hearing by learning songs of resident currawongs and kookaburras announcing dawn in our black sallee trees.

One parallel between eucalypts and my young-mom status was the "mother tree" phenomenon. In the scientific literature, some temperate species have been given the affectionate term of "mother trees" because they appear to share underground resources between adults and juveniles. A mature individual with extensive mycorrhizae (fungal associations that advantageously take in water and nutrients from the soil) may share resources with nearby juveniles. In the tropics, however, growing up near a parent tree often leads to greater likelihood of attack by predators because juveniles are easier to find, so seedlings usually have greater success germinating far away from conspecifics. Consequently, the juveniles in tropical forests are not likely to grow close to a parent, making the mother tree behavior unlikely. I did some limited work on mycorrhizae in tropical trees after my PhD was completed, and published a paper with my advisor, Joe Connell, theorizing their underground network conferred a competitive benefit for water and nutrients to certain species that had mycorrhizal partnerships. We formulated this idea because some tree species grew in dominant patches in the tropics and we hypothesized that they outcompeted other species via underground resource-sharing. Tropical forests, however, are renowned for the existence of highly diverse stands, meaning mothers and children of the same species don't usually live nearby nor do they assist their own kind through an underground network. Scientists now recognize that trees "communicate" in at least two ways: via underground connectivity as well as aboveground emission of volatile oils from the leaves, which are airborne to warn neighboring trees when defoliators are attacking.

Another mother tree concept I observed in tropical trees as well as with eucalypts was a massive reproductive effort after they suffered insect attack or disease. When adults were stressed and close to mortality, they oftentimes flowered and fruited profusely, as a last-ditch parental effort to propagate their genes into the next generation. I could usually predict which ones were close to death because they flowered like crazy, as if flashing a message to confirm their dedication to survival of the species. After massive flowering efforts, they almost always died the following season, creating a mother tree signature on the landscape which allowed accurate prediction of mortality. I felt motherly toward my gums, so it was always bittersweet when a particularly beautiful tree staged an enormous flowering event, signaling the imminent end of its life.

Despite the distractions in everyday life, I moonlighted working on canopy data, which clearly confirmed that the massive defoliation events by Christmas beetles, sawfly larvae, and a few other herbivores ultimately led to massive mortality. The insect attacks were the ultimate last straw in a chain of environmental stresses. During that decade, climate scientists announced that Australia was becoming hotter and drier, and in the outback, we witnessed those environmental extremes culminated by insect outbreaks. Soils were not changed, pollution was not part of the equation, nor had significant clearing on local farms occurred for many decades. Nothing else was glaringly obvious except a double whammy of heat and drought stimulating major insect outbreaks. With defoliation levels of up to 300 percent annual leaf area loss (meaning three flushes eaten by insects per year), the poor trees had no chance. Christmas beetles, the major culprit, thrived under rural pasture conditions and they destroyed both roots and leaves, a scenario almost like "The House That Jack Built." When farmers thinned pasture trees, less foliage remained for beetles to eat. With fewer nesting sites, the birds that otherwise ate the beetles disappeared. When adult beetles ate leaves on a fewer number of remaining gums, they ate proportionally more, so those individual trees died, and livestock huddled under the ever-shrinking canopy shade. Those sheltering sheep and cattle then saturated the soil with nitrogen via their waste, creating ideal conditions for the next generation of beetle grubs. More larvae hatched to consume a depleting supply of roots and leaves. Ultimately, the increased populations of beetles (both grubs and adults) led to the death of the last remaining trees.

I was glad our findings pinpointed the underlying cause of dieback, and that it was insects, not that charismatic canopy marsupial. What a great day for the koalas, exonerated of all guilt; these furry herbivores did not cause tree deaths as rumored by some graziers. Having had the privilege of encountering koalas throughout my arboreal activities, I felt a special bond to these vegetarians. Their fondness for leaves paralleled mine, and even more so because they not only lived and breathed in the treetops, but they also exclusively ate gum foliage, consuming approximately 400 grams (14 ounces) of leaves each day. On several occasions, I actually patted a koala on its butt, since it was so sluggish after a salad bar feast and did not find my presence threatening. Koalas are now listed as a vulnerable species by the International Union for Conservation of Nature (IUCN). The 2020 fires not only decimated its populations, but also destroyed its canopy habitat. It will require decades to restore eucalypts into adulthood for koala habitats, and successful landscape restoration will need the exclusive species that koalas prefer, including *Eucalyptus microcorys*, *E. camaldulensis*, and *E. tereticornis*. Koalas are great climbers, and I was honored to share the treetops with this largest of arboreal marsupials.

Hal and I published the results of this complex interplay of humans, drought, heat waves, landscape, livestock behavior, beetle life cycles, and ultimate tree death. Other regional scientists contributed pieces of the puzzle, including studies of soil pH and monitoring of bird population declines, but insect defoliation provided a final nail in the tree coffin. Understanding eucalypts was so important to both ecology and economics that our results were featured on national television. Hal and I also coauthored a book, dedicated to all the farmers who assisted with our research and benefited from our findings. I shared news of the dieback book publication with the in-laws, but it fell on deaf ears, despite inviting them to a big lamb roast with homemade apple pie to celebrate. Even my husband could not convince his parents otherwise. In those days, ecological research did not usually link to the economy, but the dieback results were different, and they led to solutions to restore tree cover and bird populations in rural New South Wales. The following year, Andrew and I helped start a restoration program called "A Million Trees by the Year 2000," and many graziers developed their own nurseries using local gum nuts best adapted to their landscapes. Over time, surrounding farms planted windbreaks and created seedling nurseries with irrigation and rabbit-proof fencing. Other botanists took the lead on farmland restoration, but our station played a pioneering role through active planting activities. Andrew's tree-wife used her tree dowry to good end, and my grazier husband was incredibly supportive.

As the months wore on, however, it became obvious to Andrew that his wife was a bird in a cage, mostly due to pressure from his parents. He had married a biologist and respected my passion for forests as part of our wedded life, but it was tough for him to defend me in the face of his mother's expectations. At one point, a beleaguered Andrew came home from work and said he was ready to leave the farm because he couldn't tolerate his dad's criticism of everything he did, nor his mom's negativity about everything his wife did. I jumped for joy. We would raise our loving family without in-laws as our nearest neighbors. Our euphoria was short-lived when a day later Andrew said he could not turn his back on his parents after five generations of farm inheritance, so we had to figure out how to survive under their strong thumb. I cried but resolved to be loyal. For better or worse, I had married this man whom I loved and had signed on to live in the middle of his fifteen thousand sheep surrounded by multitudes of dying canopies.

My mother-in-law continued to criticize any ongoing field research, although efforts to juggle child-rearing and household duties only allowed for a limited amount of science. We had a serious falling-out when I took the boys to America to see their other grandparents and returned to find the two rows of giant elms along our driveway trimmed (I called it butchering) by my father-in-law. He claimed it was for safety, but in my eyes it was a tree war, and he had won. I cried for days. In mourning, I walked despondently around our desolate pastures to gaze at a few struggling trees. Seeing those plucky New England peppermint trees trying to leaf again after repeated cycles of insect attack and drought reminded me not to give up. If their canopies can be so resilient, then surely I could survive one set of in-laws who didn't respect their field biologist daughter-in-law. Maybe I needed to become more strategic in moonlighting the pursuit of science? It became a game. I hid an *Ecology* journal in the middle of a *Woman's Weekly* magazine, and it looked as if I were reading recipes. I often took the boys to the university and gave them lots of toys in a wonderful cave (aka the space under my desk). While James mostly slept in his bassinet, Eddie quickly became a little naturalist in his own right, learning all the Australian bird songs played on car tapes during our commute. People took note when this tiny person enthusiastically interrupted an adult conversation saying, "Mommy, can you hear the pied currawong?" or "I think it's the black-faced cuckoo-shrike!" Thanks in part to their mom's research, both boys became intimate with biodiversity and developed a love of nature at a young age. In addition to wildlife, we had eleven dogs on the farm, but these animals were not meant for human affection. On a sheep station, dogs are primarily work animals, trained to follow their master's directions. When the dogs rounded up three hundred pregnant ewes in a paddock, it was like watching a beautifully orchestrated dance. But if a dog ventured inside the house or became too affectionate with people, it was usually shot. It was not their role to be a pet or loved by children. Eddie and James felt confused when we visited America and saw people lavish attention (and money) on four-legged animals that didn't even work to earn their keep. As young adults, both boys now have dogs as pets, so I guess they overcame their culture shock!

Another difficult issue between mother-in-law and daughter-in-law was the boys' early enthusiasm for books. At age three, Eddie accompanied me on a bus to Queensland, as part of a research team to census rain forest trees. He accompanied me on most expeditions because there was no one to mind him at home. Mom was visiting and kindly stayed home with James, making it a special one-on-one for mother and older son. While sitting for six hours, Eddie suddenly started reading *Green Eggs and Ham* by Dr. Seuss. I was gobsmacked! Maybe he had memorized it? When we got to the rain forest lodge, I handed him the dinner menu, which he also read. During one long day on a bus, with thanks to *Sesame Street* and lots of reading aloud by his mom, Eddie had figured out how to sound out letters and put them together into words. I was beaming! But when we returned to the farm, my mother-in-law scolded me. She thought it was dreadful Eddie was reading, because now he would be bored in kindergarten. And James was not far behind his brother in terms of zest for learning. They both preferred Legos and trips to the rain forest more than looking for flyblown ewes, a never-ending task for graziers to identify and immediately treat any sheep with infections of fly larvae on their butts. Still, one day Eddie announced to me that girls could never become doctors, and I realized that—despite my best efforts—he was absorbing the gender attitudes that permeated the rural culture. How could I reverse an emerging bias in the wiring of his young brain?

After the university research position terminated, I started a bed-and-breakfast to save up enough funds each year to fly the boys to visit their American family. The B and B became one of the first authentic farm stays in Australia and garnered a lot of media attention. I loved cooking three-course meals (having watched my mom cook healthy and creatively with a very modest budget throughout childhood) and amassed a huge stack of recipes for homegrown lamb. (Maybe those white recipe cards were simply another collection, like the wildflowers from youth?) I decorated a guest suite with unusual Australian crafts and woodwork, and acted as an expert nature tour guide, taking folks on four-wheel-drive adventures through the paddocks to see birds, kangaroos, and sheep. I built a nature trail through our dying trees, with a printed pamphlet, so visitors could explore local botany and ecology. Life was incredibly partitioned, almost frenzied, between housewifery, mom duties, hosting guests, and moonlighting with canopy data.

In addition to the social enrichment from B and B guests, I kept close contact with my local female friends, all tied to their farm properties as wives and mums. We hosted weekly playgroups, despite the many miles between our properties. And we commiserated, since there was no hope of creative employment given our isolation. In our district, the sheep stations were large and financially sound, but many wives wanted the challenge of an occupation outside the home. One girlfriend started an embroidery shop in our tiny town, and another an art gallery. Their enterprises barely limped along in a rural setting surrounded by treeless landscapes that discouraged tourism. The Australian government had actually paid for my graduate studies, and I still hoped to use that training for the greater good. After the boys were born, I applied for an assistant professorship at the local university. Riding on the success of demystifying dieback, I was easily the most qualified candidate in the eyes of most colleagues. But at the first interview, I was informed that a farmer's wife, and especially a young mother, could not possibly undertake a professorship. I was dismissed and dismayed. Those girlfriends bolstered my spirits by reminding me that equal opportunity did not exist in Australia. They held my hand during such a disappointing experience; most of them were used to being treated unequally, though I was still in denial about the "Sheila" concept. Their sisterhood and support remain one of my life's best treasures.

We had a party line, so about a dozen sticky beaks could overhear any phone conversations, and some of them seemed to spend the day listening for gossip. So I communicated any private thoughts to my mom via airmail letters, which took almost a month to arrive. One morning, I was making apple pie and tripping over Legos in the kitchen when the phone rang. The call was from Pennsylvania State University. Jack Schultz, who pioneered the concept of tree communication via airborne chemicals that warned other stands about defoliators, was a true inspiration. He mentored me on rain forest leaf chemistry and tracked me through publications. In those days before internet, we all airmailed copies of our latest scientific papers to colleagues, and he had followed my career from afar, as I had with his. Even as a housewife, my publication record was prolific because I had such a wealth of data from both rain and dry forest canopy research. He had noticed a hiatus and called out of the blue to inquire if I was okay. "What the hell are you doing?" he yelled into the phone. "Making apple pie and picking up Legos," I answered. He said, "Get out of the outback and come back to the treetops—you are too valuable to the scientific community to give it all up." I hung up and cried. There did not seem to be any answers, and my fate was sealed. Or so I thought.

A week later, the phone rang again. It was Williams College asking me to be a visiting professor for a semester. I was totally surprised and felt as if I had just won the lottery. With great trepidation, I told my husband about this invitation. He was not enthusiastic but shrugged and simply said, "Just get it out of your system and return to the farm." The response from the in-laws was far less accepting, and I thought fistfights might break out. They believed no wife could juggle kids plus a job. I was daunted but determined to pull it off. Preparations were overwhelming: packing the boys' toys and other essentials, leaving the kitchen in order, satisfying my mother-in-law that her son would not starve, tending the garden so it would not shrivel and die, creating lesson plans to teach college biology, finding a wardrobe that looked professional, plus renting an apartment halfway around the world long before the days of cell phone apps (or cell phones for that matter!). I packed with greatest anxiety over three long months—which is an understatement. If anyone had suggested that the boys must stay behind, I could not in good conscience have left the farm. At this stage in their young lives, their dad was up at dawn and home after dark, so they already lived a single-parent existence. But we three bonded from so much time together, especially in the absence of television or technology. Although I lost valuable years to move up the career ladder like many American counterparts, sharing exclusive home-alone time with both boys was heartfelt and transformational in many unspoken ways. We not only learned to avoid venomous snakes together, but we traveled to Australian reefs and rain forests, as well as halfway around the world to visit their American grandparents. Because the flights were so long, I often booked a wheelchair for Eddie, because I literally could not carry both boys, plus hand luggage, between gates. It was no joke to experience international flight delays and invariably miss the last leg when we were on the verge of collapse!

On the eve of our departure for my six-month sojourn as a visiting professor, Andrew disappeared to his local watering hole, the Walcha Road Pub, and never came home. In some unspoken fashion, maybe he was giving me the wings to fly free? Because our marriage ended up in a stalemate where the outback culture frowned upon my pursuing a science profession and the family farm history defined Andrew's future within the limits of a rural culture, I believe in my heart that he was reluctantly freeing his bird-in-a-cage wife to follow her dreams. At 4:00 a.m., realizing we had no ride to the airport, I called a reliable girlfriend, Nena Fay. About fifteen miles away, she was one of our nearest neighbors. She immediately rose out of bed and drove us an hour to the local airport for our propeller flight to Sydney. Andrew never said goodbye, which may have been a blessing for all of us. I breathed a tearful sigh of relief when our international jet took off over Sydney Harbor. Although the boys were only three and four years old, they loved the view and relished the airplane headphones, chattering excitedly. I was emotionally exhausted from wrangling myself to take this "intellectual asylum." What lay ahead? Would we survive in a new culture? Could I ramp up my intellect from Legos to forest succession, to engage some of the world's brightest college students? How long would it take the boys to adapt to streets with sidewalks and towns with bookstores? It was a huge gamble to think we could not only survive but thrive. We were heading to America.

⫸ New England Peppermint ⫷
(*Eucalyptus nova-anglica*)

IF THERE WERE AN EMMY AWARD for Best Performance of a Tree, surviving despite amazing odds against six-legged enemies, New England peppermint would win. It is the only tree in the world where I observed insects completely defoliating the canopy not once but three times during one season—and a few individuals still even managed to flush a fourth time. Its resilience exceeded any other canopies I ever encountered, but its numbers are significantly declining. New England peppermint suffered the highest defoliation of any species throughout field research, losing an average of 60 percent annual leaf area loss per leaf! This extraordinary herbivory spiked to over 300 percent leaf loss (meaning herbivorous insects ate the entire canopy three times) in the 1980s, leaving entire landscapes littered with white skeletons. A similar event had been observed by graziers in the late 1800s, when another series of droughts were recorded, leading to insect outbreaks on gums. Fortunately, a few individual trees in isolated pastures survived the voracious Christmas beetles and sucking aphids of the 1980s dieback, and those remnant trees will foster a new seed stock of future peppermints for the region. In the aftermath of the late twentieth-century dieback, farmers also created nurseries and land-care groups to plant natives, especially peppermints.

The story of New England peppermint embraces Australia's agricultural history. New England peppermint's name is derived from Latin: *nova* means "new" and *anglicus* means "England," referring to its occurrence in the New England district of New South Wales, a plateau of 3,000 to 4,500 feet in altitude. The region was so labeled because some early settlers planted stands of deciduous trees that change color in autumn, looking similar to the American Northeast. In true botanical form, much of New England peppermint's scientific description requires a Latin dictionary to decipher: *juvenile leaves opposite, orbiculate, cordate, glaucous; adult leaves disjunct, concolorous, falcate, lanceolate; conflorescence simple, axillary, umbellasters 7-flowered and regular, with peduncles terete or quadrangular; and fruits conical or hemispherical, pedicellate, 3–4 locular with disc raised, valves exserted, and chaff dimorphic, linear and cuboid.* In layman's terms, the juvenile leaves are round and silvery green, while the adult leaves are bluish green with long tapering shapes. Although the jargon is technical for a layperson to decipher, such level of detail allows botanists to distinguish it among the other 555 species in the genus *Eucalpytus*.

According to a review in *Science* magazine in 2018, agriculture was the reported cause of 27 percent forest loss on a global scale. In Australia, clearing of the dry forests for livestock grazing has created a landscape of isolated remnant trees, sometimes stressed beyond survival due to their exposure after the loss of surrounding stands. Less than 10 percent of the so-called New England peppermint grassy woodlands remain in 2020, due to agriculture and related human activities as per the dieback episode of the 1980s. Now existing forest fragments are protected under Australia's Environment Protection and Biodiversity Conservation Act 1999 (EPBC Act), as a critically endangered ecological community. Others growing in these patches include snow gum (*E. pauciflora*), black sallee (*E. stellulata*), mountain gum (*E. dalrympleana* subsp. *heptantha*), Blakely's red gum (*E. blakelyi*), and fuzzy box (*E. conica*). These stands house an amazing diversity of Australian natives: marsupials such as feathertail and sugar gliders, gray and red kangaroos, common brushtail and ringtail possums; foragers like common dunnart, whiptail wallaby, and brown antechinus; carnivores like dingo and red fox; microbats including Gould's wattled bat, chocolate wattled bat, long-eared bat, and southern forest bat; many birds; and a multitude of insects (many herbivorous).

The New England episode was a complex forest catastrophe, for which there was no single cause. Its impacts were synergistic, with insect outbreaks of Christmas beetles, sawfly larvae, and a few others as final stressors in the struggle of gums to survive. The New England peppermint was iconic as the face of this tree apocalypse. Fortunately, a community effort of scientists, graziers, farmers, economists, foresters, land managers, and politicians shouldered the responsibility to restore the trees. Groups of rural land-care groups sprang up around the district, collecting and planting native seeds, and our group called "A Million Trees by the Year 2000" achieved its goal a decade ahead of schedule. Peppermint was one of a dozen or more species prioritized in the restoration of the New England tablelands. With increasing climate extremes, Australia can expect more droughts, increasing heat waves, and additional challenges for the survival of forests. Fires have become more widespread, as during 2019–2020 when large portions of the New England region went up in smoke. In addition, the water table has deteriorated, with increased salinity. Are gums resilient enough to tolerate the rapid onset of climate change? Is the genetic variability of seed banks diverse enough to ensure their adaptation? Only time will tell. As an arbornaut, I keep shouting for the trees, including peppermints, as they increasingly fall victim to human carnage.

6

HITTING THE GLASS CANOPY
⋙⋘

How Strangler Figs and Tall Poppies Taught Me to Survive
as a Woman in Science

I HAVE TWO BITS OF WISDOM to share with every young girl:

1. Don't ever hesitate to be smart and strong.
2. Always nurture and support other women.

I was part of an emerging generation of women seeking equality, but afraid to admit we were leaving work for our child's doctor appointment and never daring to say no when asked to make coffee at a faculty meeting. I used to believe that to succeed meant rushing home after teaching to do laundry, cook dinner, and help with homework, yet many male colleagues stayed late at work without guilt, headed to the pub with their old-boy network, or played golf to garner a promotion. My female peers and I may have been trailblazers in field biology, but we bruised ourselves on a glass ceiling every time we reached beyond what was expected, so much so that I came to anticipate—and even worse, tolerate—the bruises. As my colleagues reminded me, the term "bruises" was probably too gentle—they were actually major cuts. Although it was a breakthrough for some of us women in science to shatter that "glass canopy," we bled out from those gashes inflicted by the broken glass, and our gender was trained to downplay the pain. Dredging up these bad memories of being a lone female tiptoeing through a male-dominated career path is still painful, but I share them in the hope that my mishaps will help readers of both genders become more informed than I was and avoid future workplace inequities.

I was never a girl who jumped into a puddle and laughed if I splashed someone; I was the one who hid behind a tree trunk when seeing a gang of boys walking home from high school. In fifth grade, the principal called me over the intercom one day in a big, booming voice. Still excruciatingly shy, I timidly raised a hand so the teacher could see. Both teacher and principal expressed great relief that I was sitting in class, devotedly finishing a math exercise. A deranged man had called my mom, claiming he had me tied up in the back seat of his car. She'd immediately hung up in tears and called the school. There I was, quietly doing division and multiplication. I had recently been featured in the local newspaper for composing a symphony called *Hoffman Air*, named for the school and performed by its orchestra. Although I loved nature, music was another passion. From my ordinary weekly piano lessons, I somehow conjured a tune in my head that I managed to translate into a musical score for the entire school orchestra. Even though I was too shy to ever perform in a piano recital, I learned how to write music, which reminded me of birdsong. It was a thrilling accomplishment for the class nature lover, but publicity obviously had its drawbacks. Mom came to school and hugged me— deliriously happy I had not been kidnapped. The local police asked me to carry a notebook for the next few weeks and record anything suspicious. That was crime-solving in the 1960s—no cell phone, text messages, photos on Facebook, or public lists of child molesters that could easily be checked, just a small notebook and pencil in a girl's lunch box.

Fortunately, the threat never materialized. But it sure made Mom anxious, and it gave me the absolute creeps. I almost became afraid of my own shadow and retreated into the safe cocoon of my bedroom laboratory to gaze at dried flowers. Mom's small-town wisdom considered it best to have your name in the news at birth and at death but never in between. During my childhood, the notions of modesty and small-town values were blindly mixed in with good old-fashioned gender bias, meaning that women were reminded not to brag, shout, or exclaim about their accomplishments. Mom gave me the only advice she ever knew, and it came back to haunt me several times as a grown-up. In contrast, as Laurel Thatcher Ulrich brilliantly said, "Well-behaved women seldom make history."

Although I still get flashbacks about the childhood kidnapping episode, I could not stay surrounded forever by the innocence of dried collections and bird eggs. Looking back, I only wish I had become street-smart at a younger age. Later in life, conducting fieldwork in the Australian jungles, I learned to admire the fig trees that dominated many tropical forest canopies and dispersed their luscious fruits throughout the entire forest. Numbering over eight hundred species, the genus *Ficus* is part of the family Moraceae, and serves as the base of many tropical food chains. In addition to providing food and shelter to thousands (maybe millions?) of species, figs offer spiritual sanctuary for several billion people in India, Africa, and Asia. One of the best examples is the Bo tree (*Ficus religiosa*), native to India and Southeast Asia, thought to be the canopy under which Buddha received enlightenment. If I am ever reincarnated as a plant, I want to become a fig because they are not only strategic and successful, but also altruistic, feeding their fruits to an entire ecosystem. Of those eight hundred fig species, however, a few exhibit adult behaviors that are less benevolent. In Australia, one of those more insidious figs, and one I frequently climbed, was Watkins' fig (*Ficus watkinsiana*), a member of the subgenus called *Urostigma* that includes strangler figs, or banyans. They have evolved the most extraordinary survival strategy of any tropical trees, bar none.

Stranglers begin life at the top and then grow down, similar to a vine in structure but then hardening as a successful tree, having already secured a space in the sunlight. Figbirds eat strangler fruits, then excrete the seeds on a branch, generally high in the canopy. With easy access to abundant light and water at that height, they germinate rapidly, faster than any counterpart seedlings on the dark forest floor struggling to grow upward into the overstory. Strangler cotyledons receive abundant sunlight, so the fast-growing seedling quickly extends aerial roots downward, reaching the soil. The fig is already actively photosynthesizing with its foliage strategically situated in the sunlight, offering a burst of energy so its roots can take in a generous share of water and nutrients, enabling rapid growth. In this unique top-down growth, stranglers encircle their host and often (but not always) suffocate them. This part of their life cycle is less admirable, and I would never advise women to strangle their competitors, but this initial strategy to secure a place in the sun before putting roots into the soil illustrates an admirable success story. And still, even the strangler's deadly embrace has a silver lining. Research has shown that stranglers keep their host upright during storms, actually reducing treefall mortality. Figs can teach important lessons to women trying to overcome hurdles in the workplace—follow a ficulneus, or fig-like, strategy to be innovative, nurture others, tap wisely into resources, and rise above the rest.

My own personal growth resembled the life of a conventional tree seedling that germinates on the forest floor—I did not think about the importance of strategically seeking the essentials (light and water in the case of seedlings) to jump-start career advancement. I didn't think about the absence of female mentors. Throughout high school and college, I was truly a wallflower and did not usually speak up. Over my eleven years in Australia, I experienced more sexual advances by male colleagues during fieldwork than can be counted on both hands (or at least that is all I am willing to acknowledge). It just seemed to be part of the landscape, in the absence of senior women to mentor young students like me and years before any of the #MeToo reporting structure. Fortunately, I got surprisingly good at wiggling out of such precarious propositions. Today, an astonishing majority of female scientists still report sexual harassment during fieldwork. From six hundred surveyed, the anthropologist Kate Clancy at the University of Illinois reported over 70 percent were victimized, sometimes thousands of miles away from their campuses that boasted strong harassment policies. Throughout a field biology career, I confronted a few detractors who may have deserved a strangler fig response from their female colleagues, in particular, several male bosses whose bullying seemed to be their nature. Like a frog placed in a pot as the water gradually heats up, I tolerated this behavior as part of the job. I really didn't maximize available resources or strategize about how to succeed nearly as effectively as a strangler fig.

When I arrived back in America as a visiting professor, I felt pretty insecure. I saw myself as a farmer's wife with two unruly sons, and naively accepted the salary offered without argument. It never occurred to me to negotiate, and as I later learned, women tend to be less effective than men in salary negotiations. Current studies show women end up with an average of 29 percent lower retirement income than men. Based on my starting salary, the boys were eligible for free lunch at school, which was both humbling and humiliating. I did not let them enroll in the program, for fear they would be targeted; after all, most of their classmates had prestigious professorial family pedigrees, not a rural outback heritage. So I faithfully made healthy lunches every day, which they often brought home untouched. When asked why, they admitted the other kids goaded them to talk nonstop during lunch, laughing at their kooky accents. But they soon lost those Australian accents and integrated well, except perhaps the rude shock of experiencing their first cold snow. Given the severity of western Massachusetts winters, our chilly rented apartment was a wake-up call to empathize with struggling families who weren't so lucky. I had the good fortune to inherit a few gold coins from a great uncle, which paid our heating bills during the first year. Despite sparse furnishings, we had one infamous piece of furniture under the stairwell, called the "time-out chair." When someone was naughty, he (or she) was relegated to that chair. The boys loved it when once (and only once!) I uttered a swear word and they assigned me to the time-out chair for thirty minutes.

As a visiting professor, I had many successes—grants, good teaching reviews, pioneering canopy discoveries made in temperate forests with my undergraduate students, lucrative graduate scholarships for all of them as a result, and great publicity for the college. Nonetheless, those previous eleven years in Australia had infected me with the notion that women were second-class citizens, which undermined all the self-worth I had built up over the years. What if the boys got sick and I missed teaching a lab? Could I juggle lesson plans with household chores? Could the children transition from their outback upbringing with its poisonous snakes and wildfires into an urban setting where different threats prevailed? I decided they both needed to get street-smart in a hurry so started by introducing them to the public playground with its sophisticated wooden play area of tunnels, swings, and slides. James got inside a wooden maze and started wailing because he had never before experienced such equipment. Both boys earned their peers' respect soon enough, however, thanks to a knowledge of natural history. Eddie had a memorable first-grade birthday party in the college forest, where my biology students climbed into the treetops and candy magically rained down from above. James attended the college preschool and was dubbed a local hero when he alerted his teacher that the provost's son had eaten deadly nightshade berries on the playground. I got a call from the day-care director asking if he was able to correctly identify plants. I answered, "Absolutely, yes," and so they rushed the other child off to have his stomach pumped. Nightshade was immediately removed from the playground. On weekends, we enjoyed many hours of activities available for free in nature: birdwatching, counting spittle bugs, and carving forts in backyard bushes. What a change—no gum trees, no brown snakes, no blowflies, but lots of new wildlife such as birch and oaks, warblers, squirrels, and the roadside wildflowers of my childhood. Truly, it was a kids' paradise to frolic in New England forests.

Despite the joys of motherhood, which included sharing lots of playtime with my young explorers, I felt vulnerable as a single parent with no tenure or other safety net. It was tough, but I was glad to be back in the science arena using knowledge gained from research half a world away, and relieved the boys were now in a place where both genders seemed to be respected more equally. I jumped out of bed each day relishing challenges that required brain cells, not just home economics, as was the case for a farmer's wife. After the first semester, both teaching reviews by students and my publication rates were excellent, so the college renewed my visiting professor status. I gathered up a big dose of courage and shyly approached the dean of the faculty, explaining that I would love to renew the contract but could not survive on its current salary. He seemed aghast and admitted I had been severely underpaid, immediately doubling the amount. Oh joy! Now we could pay for heating, and our household budget exceeded the eligibility for free lunch!

The cultural transition from Australia to America was probably tougher on me than the boys. After suffering cruel rejection from that first job interview in Australia when the search committee claimed a farmer's wife with toddlers couldn't possibly qualify as a professor, I felt grateful that someone had thrown me a crumb. In my former part-time job for the EPA during my graduate year at Duke, I was happy to make the coffee for all the male engineers. On most field expeditions, I was invariably the organizer of food and logistics, having occupied a similar role on the farm despite the fact that I was never authorized to use the checkbook. I got surprisingly good at smiling, nodding, and acting gracious in the role of second fiddle to menfolk. As my career advanced, the school of hard knocks left me bruised, but I was slowly becoming wiser, bolder, and more strategic. However, with career success came another insidious danger faced by many females climbing a career ladder in the late twentieth-century workplace—"tall poppy" syndrome. In Australian English, "tall poppy" was a slang term originally aimed at celebrities, accomplished businessmen, and the wealthy elite to denigrate their success, but it subsequently permeated the workplace. This Aussie phrase reflected a cultural tendency to encourage mediocrity by cutting accomplished people down to smaller size. At the time I was building a career, men in the workplace were increasingly threatened by strong females seeking equality. There were a few wonderful examples of female peers who seamlessly worked their way to the top, but usually they had supportive spouses or that exceedingly rare species: an empathetic male boss who encouraged and promoted them. I greatly admired the successes of such women. But many of us simply stumbled haphazardly over gender hurdles. Field biology remained a male-dominated profession throughout much of my tumultuous career. I tried to ignore all the unbalanced ratios and published furiously, all the while minding the children and kitchen. I never dared speak up about workplace inequities, nor did most female peers. We considered ourselves lucky to have a desk and were terrified of losing it.

When I started my visiting professorship in Massachusetts, I reached out to someone I truly admired, sending a fan letter to Jill Ker Conway, the president of Smith College, and thanked her for a best-selling book she wrote called *The Road from Coorain*. Mom had mailed it to me while I lived in Australia. Jill grew up on a sheep farm close to ours, and her memoir of a girl trying to compete in a male-dominated academic world rang all too familiar. After reading the first two chapters, Andrew ridiculed the book, but my outback girlfriends loved it. I briefly explained to Jill how my career opportunities had been sidelined for seven years as an Australian farmer's wife, but I was thrilled to be a visiting professor in America for a semester. Much to my amazement, Jill wrote me back with heartfelt advice. She stated to never return to Australia but seek intellectual asylum in the United States, and she followed up by sending the name of her lawyer, suggesting I file for divorce. This distinguished woman reached out to mentor a fellow female, and I am forever grateful for her personal advice.

Not surprisingly, the phone calls between Australia and Massachusetts were not amicable from the start. I can only guess the boys' father had gambled on the notion we would return before the end of first semester, as his parents had predicted. Those loyal rural girlfriends reported my mother-in-law was scouring the countryside for a suitable replacement for her son's unconventional scientist wife. I couldn't blame her—I didn't represent the ideal daughter-in-law she had so wished for. I had a consultation with Jill Conway's attorney, who explained I needed to seek legal representation in Australia. It was bittersweet when a rural Australian neighbor called out of the blue and offered to represent me in a divorce filing because he and the entire outback community believed our marital situation was fueled by a mother-in-law's interference. In a sense, Andrew and I were victims of a culture.

As the wife of a grazier, I had the right to ask for half the farm in divorce proceedings—representing millions of dollars in land value and allowing me to live comfortably for a lifetime. Ethically, I did not wish to do so, because a working farm needs to remain whole, without fragmentation from family feuds. That was a contentious time, juggling both job and kids plus the anger exploding via long-distance telephone calls. It was a standoff: Andrew unwilling to come to America, and I not about to break the teaching contract. Our love for each other had been eroded by a culture. Despite the joy of teaching, I missed those morning kookaburra songs and kitchen-window vistas of sheep grazing in a landscape of eucalypts. But I was determined to try a hand at science after a lifetime of training, and even more resolute to provide the boys with an American education where both genders were more equally respected, even if it meant facing the uncertainty of single parenting. I figured a few years of American public school education would enable both boys to return someday and run the farm using new technologies. Without good schooling, their lives would be forever limited to sheep and more sheep.

I loved teaching and achieved an element of national fame by building North America's first canopy walkway in the college research forest, based on the earlier Queensland aerial trail. Because of my canopy reputation, I had quickly integrated into the arborist community in Massachusetts. Two of them had encouraged me to write a grant, enabling the three of us to build the walkway spanning several red oaks, with two platforms and one simple bridge based on that original Australian skywalk. Now the temperate canopy was opened up to exploration. Conducting observations at a height of seventy-five feet, one arbornaut student discovered that southern flying squirrels (*Glaucomys volans*) served as a natural pest control, feeding on a notorious New England oak defoliator, gypsy moths (*Lymantria dispar dispar*). The United States Department of Agriculture had spent millions of dollars researching gypsy moth outbreaks, but most of their fieldwork had been limited to ground level. Like my own experience in Australian rain forests, these students made original research discoveries by studying the whole tree and not just its big toe.

The divorce was finalized by telephone. The Australian attorney executed the paperwork in short order, and both Andrew and I were somewhat relieved. His mom was doubtless dancing a jig. It was bittersweet because I genuinely loved him, yet our inability to integrate sheep and trees into one life seemed irreconcilable. The college renewed my contract for another year, and during the January term I took students to Florida to study subtropical ecosystems, with my boys in tow. At age eight, Eddie was a great birding instructor, and it was good for students to witness their professor juggling kids and career. Even though field expeditions with kids plus students wore me out, I was aware how few role models existed for women in STEM in the early 1990s, and this notion kept me energized. Most other biology faculty had stay-at-home wives. I was also fortunate to have parents only five hours away by car who frequently came for long weekends so I could grade papers or write lectures. Occasionally, they stayed with the boys during longer field expeditions, incredibly supportive but always anxious about their daughter's remote destinations.

A global research expedition to Africa was my first exhilarating exposure to a third and innovative new canopy access technique, this one of inflatables and called "Radeau des Cimes" (raft on the roof of the world), designed by a team of French botanists. A tiny ad in the weekly issue of the journal *Science* caught my attention: "WANTED— field scientists for hot air expedition in Cameroon, Africa." Intrigued, I applied. Thinking back to Australia where I was either excluded or very much outnumbered as a female, a sixth sense forewarned me to apply as "M. Lowman." I was accepted as their field specialist for insect-plant interactions; no one asked about my gender. Technically, this expedition utilized a dirigible slightly different from a hot-air balloon because it flew using a gas contained inside the frame to make the contraption lighter than air. (In contrast, a hot-air balloon flies by heating hydrogen in its fabric.) When airborne, the dirigible was capable of towing an inflatable raft that could be shifted between treetop locations to serve as an aerial base camp. The lightness of the inflatable raft enabled it to be tethered on top of the uppermost branches without breakage and easily towed to a new location every few days.

Additionally, a triangular pie-shaped section of the round raft could be towed by the airship each morning (before the prevailing winds came up) to sample multiple neighboring trees during one time frame. All these inflatables—dirigible, raft, and sled—offered unique access to the uppermost canopy and, let's face it, were truly fun!

This new adventure into the jungles of Africa was beyond comprehension. How could such a vast landmass, otherwise known as the Biafran Congo rain forest basin, have remain so unexplored and devoid of scientific discovery in the late twentieth century? I found a tiny dot on the map for Réserve de Campo, which was our destination in southwestern Cameroon, considered one of the most diverse regions of equatorial forest. I looked forward to comparing the African and Australian canopies.

Francis Hallé, the French botanist and expert on tree architecture, was the creative mastermind behind this balloon invention that soared over the forest, plus its supplemental inflatable raft anchored to the upper crowns. The French nation was extremely proud of its scientists who had designed such innovative treetop access tools, collectively called Opération Canopée, and it was widely featured on national television. When I flew through Paris, the customs officers cheered when they saw Radeau des Cimes stickers decorating my luggage. After many flights, I landed in Douala, the capital of Cameroon, which was under a military regime. Six of us had flown in as the American contingent and squeezed into a four-wheel-drive vehicle, with me in the front seat as a female distraction, intended to disarm the military checkpoints. We drove for many hours on narrow dirt roads, arriving at night. I was relieved to sneak into camp under the guise of darkness, already anticipating some disappointment in my colleagues because their new participant, M. Lowman, was not Mark or Michael. Suffice it to say the balloon team seemed less than happy to see me alight from the vehicle. I later realized I was the only female on a fifty-person team during that entire month. In hindsight, I was fairly naive to traipse off to the jungles of Africa with forty-nine men I hardly knew. But over the course of the expedition, the male scientists came to respect me, probably because I was a competent climber. Nonetheless, I endured many moments of doubt due to gender—could I succeed in this male-dominated world of field biology despite the fact I was one of its foremost contemporary pioneers?

Perhaps as an initial test of female endurance, I was assigned the edge hammock location, which also happened to hang directly above the hole of a deadly venomous Gaboon viper (*Bitis gabonica*). This snake was not only big and ugly, but there was no known antidote to its bite. Even worse, I soon learned those male colleagues repeatedly hauled the slithery beast out of its underground lair with a noose to take photographs, which probably left it pretty cranky. I gingerly stepped around its hole on frequent journeys to the outhouse, noting with envy how other male colleagues relieved themselves directly off the deck. We late arrivals strung up our hammocks in the dark, bottoms bumping due to the cramped sleeping quarters. The base camp was otherwise ingenious, comprised of four makeshift open-air structures with thatched roofs and raised wooden floors suspended on poles about three feet above the ground. One long structure served as the hammock hut; another was a laboratory with three state-of-the-art computers run by a generator; a third, an eating hut; and farther away, an ablutions hut with five shower stalls gravity-fed by a water barrel on the roof, plus four toilet pits. The first night was tough—hot and humid, plus surround-sound snoring of forty-nine male scientists. But by the end of two weeks, I had made lifelong friends with all the team and not only knew their snoring habits but also their life stories, usually confessed after several rounds of whiskey. Francis Hallé, our intrepid leader, had imported loaves of French bread plus a case of his favorite scotch to share at the end of each exhaustive climbing day. As a funded participant on this expedition, I was allowed to bring two assistants and selected two close friends and colleagues, both male because there really were no female equivalents. Bruce Rinker was an enthusiastic educator who planned to write about the expedition for K–12 audiences, and Mark Moffett was an entomologist and expert photographer who could document the biodiversity, especially its tiniest six-legged residents. We three comprised the plant-insect interaction team.

The night sounds of the African jungle were better than any Hollywood soundtrack: nightjars, frogs, cicadas, and predawn hornbills interspersed with animated French voices as the team prepared for inflation and launch. The airship was always launched at sunrise to allow at least four or five hours of aerial activities, since prevailing winds threatened the safety of flying after about 10:00 a.m. Despite our late-night arrival, Mark jumped out of his hammock before dawn, enthusiastic to photograph the balloon launch. He shrieked, and I looked down to see blood spurting out of two pinhole marks in his foot. We stared at each other in horror—had the Gaboon viper struck him? Mark wiped off the blood and grabbed his camera, not wanting to miss a photo opportunity. Scientists are an endangered breed who will risk life and limb for a data point, a photograph, or a collection. Mark was no exception, and true to form, he coveted the dawn launch photos more than bandaging a venomous snakebite. Go figure! Only several hours later in broad daylight did we notice his insect calipers wedged in the hut's floorboards, sharp pointers sticking upward, which had stabbed him in the darkness. He was delirious with joy to know his death by snakebite was not imminent.

The dirigible was unfurled and inflated every single day. If it were left exposed, birds, attracted to the colors like a giant flower, might peck at its stripes. At 4:00 a.m., the flame was ignited that warmed the air inside the frame. Liftoff! The gondola under the dirigible seated six scientists for daily aerial surveillance, and the dirigible occasionally towed the sled, which held three scientists, across multiple tree crowns. Every five days, the raft was shifted to a new aerial location, to allow sampling in different tree crowns. Competition was fierce for seats aloft in the balloon and sled. Our plant-insect interaction team was not scheduled for sled time until the very last day, so we climbed many tall trees throughout the week using traditional ropes and harnesses. Throughout the day or night, up to a dozen researchers climbed into the larger raft, and my team of three collected nocturnal herbivores feasting under the cover of darkness. Working in the raft led to several extremely wet sampling sessions with the sudden onset of torrential rains. We frequently got so absolutely soggy that one night we brought up a bottle of champagne and made a soaking-wet toast to the canopy. Throughout the week, we eyed the weather and anxiously hoped that winds or thunderstorms would not cancel our scheduled sled sampling for the last day. We awoke around 3:00 a.m., exhilarated in anticipation of this incredible journey. Grabbing nets, plastic bags, clippers, notebook, and pens, I gulped a modest amount of water (not too much so my bladder would endure the morning aloft) and literally ran down the trail to the sound of a tiny generator inflating the balloon. The inflation process was akin to tribal chiefs undertaking a special ceremony, with all the French engineers and design team encircling the expanding plastic and lovingly massaging it into full capacity. Once inflated, balloon pilot Dany leaped into the driver's seat and lit the simple flame that heated the balloon's interior. As it warmed, the entire contraption lifted a few inches off the ground, then a foot, and suddenly it was airborne, dragging the sled underneath. The three of us quickly jumped into the triangular contraption, and away we sailed. This first flight skimming over the canopy in an inflatable sled transformed my perceptions of tropical forests forever. The African jungle looked like a vast field of gigantic broccoli, with its gnarly tops of foliage at many different heights and green hues. Up at least 150 feet, we approached the top of a cardboard tree (*Pycnanthus angolensis*) and got within a foot of its characteristic insect damage peppering all the leaves. Dany hovered the contraption delicately over the crown and we jumped into action—sweeping insects with our nets and clipping foliage with our pruners (very carefully so we didn't cut a rope or, even worse, slice the inflatable). After a rapid sampling frenzy of about ten minutes, we floated off to a second species called veludo (*Dialium pachyphyllum*) to repeat the collection process. After almost two hours, we had sampled a record number of fifteen trees by gliding over their crowns and clipping a few uppermost branches, as well as netting insects in the leaves. Using ropes, this sampling assignment would have taken the better part of a week. Throughout the flight, the pilot was in total control of our safety and, ultimately, our lives; one crash into a trunk or one smash into a large ant nest could have been a mortal disaster.

Back on the ground, I spent all afternoon sorting leaves and measuring herbivory on a high-powered computer. Pierre, a French technician, had a scanner program in the computer hut, so I measured leaf areas in the deepest part of Africa using a 60 megabyte computer and, at the time, state-of-the-art digital software. This was more advanced than any prior fieldwork over the past fifteen years throughout Australia, New Zealand, Indonesia, Scotland, and American temperate forests. The French team was not the only group to bring sophisticated equipment. The team from the Max Planck Institute in Germany brought 1,300 kilos of highly technical equipment to measure foliage respiration, but theirs had been detained by the military in Douala. I felt grateful to rely on the simplicity of clippers, plastic bags, waterproof pens, and tape measures for sampling essentials.

As the only female scientist in camp, there were a few disadvantages. First, my underwear kept disappearing, and by the end of the trip, I only had one pair left. Later, some of the scientists observed the locals quietly snatching them from the clothesline whenever I rinsed out any gear and informed me that female lingerie was coveted in the nearby Pygmy village. A second incident involved showering—whenever I entered a cubicle, several local workers invariably jumped on the roof to tinker with the water tank, giving them a perfect view of my anatomy. Life in base camp was not without humor! Another minor snafu was unknowingly bringing a few legacies of motherhood among the field gear. I forgot the local bank had given me two lollipops for the boys before departure until I saw swarms of ants raiding my backpack. In frustration, I tore open many zippered pockets to find the sugary remains of an ant feeding frenzy. Even with sealed wrapping, the candy was detected and devoured by these aggressive arthropods.

Radeau des Cimes earns top marks from me as the best canopy tool in existence at that time. This third method truly embraces the collaborative potential of scientific research because everyone shares in the thrill of discovery as a team. Unfortunately, the cost to deploy these inflatable gadgets with a full team for a month to a country like Cameroon is close to a million dollars, an excessive amount in the underfunded world of tropical field biology. As a result, the equipment spends more time in mothballs than in the treetops. But the Cameroon expedition was a huge success for the herbivory team, and our data was integrated into a comprehensive volume about African rain forests, edited by Francis Hallé. My global surveys of herbivory were expanded by this new data point in Africa as I continued to calculate how much insect damage different forests can sustain. We piloted a new insect-sampling technique called "sled sweeping" and collected a range of two to thirty-two insects per net sweep in the uppermost crown, a region not previously explored. African canopy leaves, sampled for the first time ever, ranged from 0 percent to 64 percent leaf surface area eaten for *Dialium pachyphyllum*, and 0 percent to 16 percent for *Pycnanthus angolensis*, but with absolutely no insect damage for most (but not all) leaves of *Alstonia boonei*. (I have a personal quest to find at least one tree species with absolutely no insect attack. Such a species might prove an amazing medicinal elixir given the apparent resilience of its toxins!) Our results averaging 15 to 30 percent foliage loss by herbivores in Africa paralleled Australia, suggesting tropical trees on at least two continents tolerated higher amounts of insect damage than previously measured by scientists restricted to the forest floor. As with all field biology, more years of measurements in Africa are necessary to confirm these preliminary findings.

I was the only team member who returned to Cameroon and collaborated with the sole local botanist from our expedition. The late Bernard Nkongmeneck desperately wanted to partner with international scientists after hosting them, so he and I wrote a National Geographic grant to survey local epiphytes and educate villagers about farming them for European markets. Bernard lamented that many tropical forests were increasingly logged by outsiders, who removed the timber but simply burned the crown remnants. We both recognized the potential value of those epiphytes so piloted a program for orchid and fern farming, providing villagers with a more sustainable income from selling air plants than timber. We published a list of Cameroon's epiphytes in the scientific literature and piloted canopy farming by using pulleys to hoist woven mats with epiphytes into the treetops to grow

My early interest in wildflowers led to second prize at the 1964 New York State science fair when I was eleven years old. (Photograph by Alice Lowman, 1964)

DDT is endangering many birds because it prevents calcium from developing in the shells. The egg on the left, dated 1860 is .019" thick, the one on the right, 1970, is .011" thick.
Decreasing at this rate, in 100 years the shells will be thinner than a piece of paper and won't hatch.

In 1970, I discovered my great-great-uncle's birds' egg collection, including a one-hundred-year-old chicken egg. Comparing it to a modern egg helped me recognize how DDT impacted shell thickness and species survival. (Photograph by Meg Lowman, 2020)

1860 - .019 inches thick   1970 - .011 inches thick

Musk Mallow

AREA: New England to Maryland west to Nebraska
COLOR: light purple
HEIGHT: up to 2 feet
BLOOMING SEASON: June-July
REMARKS: Flowers are clustered at the top of the plant.

This pressed plant comes from my childhood wildflower collection, picked along roadsides. (Photograph by Meg Lowman, 1963)

My family were great supporters of my interest in nature and we often enjoyed picnics near our home in Elmira, New York. (Photograph by Alice Lowman, 1962)

As a teen I was a camper and then a counselor at Burgundy Wildlife Camp in West Virginia. Here, I'm sixteen and surrounded by walking ferns, a rare resident of the forest floor. (Photograph by Bob Kluttz, 1969)

Birch trees that grew high along elevational gradients in the Scottish Highlands were shorter and scrawnier than the trees at the bottom. (Photograph by Meg Lowman, 1978)

This tough, gnarly birch tree grew at the very top of Ben Tee in the Scottish Highlands. (Photograph by Meg Lowman, 1978)

As with many scientific discoveries, I learned quite by accident that many insect herbivores feed at night, not in the day. To study herbivory as it happens, arbornauts have to climb up into the canopy at night. In this photo, the next-generation arbornaut Dr. Lily Leahy night-climbs in Queensland. (Photograph by Steve Pearce, the Tree Projects, 2019)

From 1977 to 1978, I studied birch trees in the Scottish Highlands at Aberdeen University. I camped so I could easily hike to my tree stands and measure birch phenology. (Photograph by Meg Lowman, 1978)

In January 1979 I made my debut as an arbornaut, climbing a coachwood (*Ceratopetalum apetalum*) in the Royal National Park near Sydney, Australia. (Photograph by Hugh Caffey, 1979)

I sewed my first climbing harness in December 1978, using seat belt webbing and borrowed hardware from the Sydney University caving club. (Photograph by Hugh Caffey, 1979)

Originally my field data were written in pencil on waterproof paper and later transcribed into a database for statistical analysis with one of the first Hewlett-Packard handheld analytical devices. So primitive! (Photograph by Meg Lowman, 2020)

Before I used my gear in the rain forest for the first time, I tested my harness on a tree on the Botany Department lawn at Sydney University. (Photograph by Hugh Caffey, 1979)

With a simple waterproof felt-tipped pen, I numbered leaves and made monthly observations of their fates. These leaves were located more than one hundred feet high in the canopy. In the first eighteen months of research, I numbered 4,183 leaves; when I finished my field work, that figure had doubled.
(Photograph by Meg Lowman, 1979)

These Australian stinging tree leaves have just emerged, so they as yet show no damage from the voracious giant stinging tree beetle.
(Photograph by Meg Lowman, 1979)

Just a few weeks into their life cycle, these giant stinging tree leaves exhibit the quilted feeding patterns of the giant stinging tree beetle.
(Photograph by Meg Lowman, 1979)

I studied insect foraging amid the simple vegetation on One Tree Island in the Great Barrier Reef. It was quite a change from the dense, dark, humid rain forests, but an excellent place to observe caterpillar feeding on *Argusia* shrubs. (Photograph by Meg Lowman, 1981)

After completing my PhD in the Australian rain forests, I moved to the outback to solve the mystery behind a major gum tree dieback. (Photograph by Meg Lowman, 1984)

To safely measure gum leaves each month throughout pregnancy, I used a cherry picker. What a luxurious and gentle ride! (Photograph by Meg Lowman, 1985)

Walking sticks are usually solitary, but one individual can devour a lot of tropical canopy foliage in just one night. They camouflage beautifully with their sticklike figures.
(Photograph by Meg Lowman, 1979)

After six years of observing gum trees and measuring their decline, I traced the cause of the dieback to this otherwise ordinary chrysomelid beetle.
(Photograph by Meg Lowman, 1985)

One of the initial suspects in the eucalypt tree dieback was the beloved koala, who relies on the gum trees' leaves for food.
(Photograph by Meg Lowman, 1984)

Once Eddie was born, I carried him with me on my leaf observations. Here: one of the few remaining healthy stands near our house stretches more than fifty feet high.

(Photograph by Hal Heatwole, 1983)

I now look for banyan (fig) trees around the world. Their top-down growth patterns are a source of inspiration and innovation.

(Photograph by Meg Lowman, 1984)

In 1994, on the Radeau des Cimes expedition to Cameroon in West Africa, we conducted surveys to measure herbivory and biodiversity. This dirigible tows a raft fifty-nine feet in diameter that holds eighteen scientists. Tethered in the treetop, it provided easy access to the canopy for repeat sampling.
(Photograph by Meg Lowman, 1997)

Our research team in Cameroon used this inflatable raft as a "base camp" in the canopy. This was really useful for finding nocturnal insects; night is the preferred time for many arthropods to visit the aerial salad bar.
(Photograph by Meg Lowman, 1997)

As a field scientist and a single mom, I brought my sons into my work. During a citizen-science expedition in the Amazon, Eddie and James helped me measure a philodendron; their leaves average twenty square feet—the biggest I have ever seen.
(Photograph by DC Randle, 1995)

Built in the Peruvian Amazon in 1994, the walkway at the Amazon Center for Tropical Studies annually hosts dozens of teams of citizen scientists 125 feet up in the canopy.
(Photograph by DC Randle, 2016)

This simple tree ladder scales the trunk of a vedippala (*Cullenia exarillata*) tree in the Kalakad Mundanthurai Tiger Reserve in India's Western Ghats mountains. Soubadra Devy and T Ganesh used this for their early research.
(Photograph by Meg Lowman, 2006)

Tea plantations encourage the removal of native trees, causing tigers, elephants, and other animals (including countless birds and insects!) to lose the canopies necessary for their survival. (Photograph by Meg Lowman, 2006)

Farmers hire guards to protect their crops from elephants. The guards watch over the fields from tree huts like this one at the edge of the Nagarahole National Park in India's Western Ghats. (Photograph by Meg Lowman, 2008)

I was part of the team put together by the Cockrell family and the Habitat Foundation to design this stunning ribbon bridge in the tropical primary rain forest in Penang, Malaysia.
(Photograph by Alan Tan via drone, 2018)

This treetop walk, called Langur Way, is just over 450 feet long. In 2017, we hosted a whole-forest BioBlitz here, giving 117 participants a monkey's-eye view of the dipterocarp forest canopy and its inhabitants. The park draws about two million tourists in a given year.
(Photograph by Meg Lowman, 2017)

My friend Tim Kover and I climbed this red meranti tree to greet the foggy dawn and test our climbing instruction site for Malaysian university students and researchers. This tree is now part of the Habitat Penang Hill, where the next generation of arbornauts trains.

(Photograph by Alan Tan, 2017)

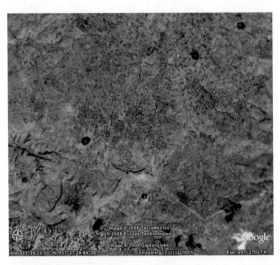

This image from Google Earth shows that more than 95 percent of northern Ethiopia's original forests have been cleared; the last tree stands, under the stewardship of the Ethiopian Orthodox Tewahedo Church, are called church forests. (Photograph by Google Earth, 2008)

Stone walls protect the Zhara church forest, which supports all of the native biodiversity and provides a spiritual santuary. (Photograph by Meg Lowman, 2016)

A close-up of the Ethiopian church forest at Bitsawit Mariam, taken via drone. Livestock grazing and firewood collection are threatening the integrity of this forest patch, but the priests have a mission to save it. (Photograph by Kieran Dodds, 2018)

The National Science Foundation funded an undergraduate research program to train students to be arbornauts. Here, Snousha Glaude and Rebecca Tripp work on their skills as Rebecca vaults from her wheelchair into the temperate-forest oak canopy of Kansas. (Photograph by Meg Lowman, 2017)

Too small to see without a microscope, tardigrades, or water bears, might be the most common resident of forest canopies. Student arbornauts collected more than forty thousand specimens of water bears over five summer field projects. (Photograph by Randy Miller, 2018)

One of the great joys of teaching? Watching your students surpass you. Here, Wendy Baxter demonstrates her mastery of arbornaut techniques on a California giant redwood, climbing hundreds of feet higher than her mentor.

(Photograph by Anthony Ambrose, 2018)

into marketable plants. We also held epiphyte identification classes in a local village after Bernard cleverly brought beer to the lectures so the villagers would be persuaded to attend. And they did! Sadly, Bernard passed away unexpectedly before we were able to ramp up epiphyte farming in Cameroon. This idea is ready to launch, simply awaiting the right collaboration of local and global botanists, but as with many international conservation projects, building local trust is critical for success.

Despite the rough conditions of an African expedition without flush toilets, real beds, pillows, potable water from a tap, or hot showers, I came back in one piece, much to the relief of my family. But back at the office, an unfortunate chain of events on campus was starting to build up into dysfunction. A new director of environmental studies had been hired before I left, but soon after, some students and staff complained about some of his interactions with them. It wasn't long before several women in the department were terminated by this director, despite their outstanding records. I had an especially successful semester, but as a single mom felt very vulnerable about a boss who did not seem to promote excellence among his staff. One day this new boss knocked at my front door, which itself was strange behavior, and announced I'd see a salary cut for the upcoming year. He offered an opaque explanation about a budget shortfall. It didn't really make sense to me, especially because I had fulfilled every responsibility and then some. That year the college achieved national recognition for my canopy walkway in the college forest, the first such structure in all of North America, plus grants and successful student scholarships, so I had anticipated a raise, not a cut. This appeared to be a classic case of good ol' Australian tall poppy syndrome, with its underlying aim: fail to reward excellence but encourage mediocrity. Two colleagues from the university where he had formerly worked confided parallel problems experienced by women who had worked with this man at their campus. Later on, I would hear that a colleague at Stanford University felt baselessly demeaned by this individual after he relocated to an organization in California.

As the sole supporter of two young children, I was overly anxious about working under these uncertain conditions, especially when it was impossible to collect adequate child support from halfway around the world. The tall poppy syndrome was truly humiliating; it really hurts to be punished for doing well. Even in America, gender bias still produced all sorts of land mines, and insidious bias was especially tough to spot in the complexities of office politics. For me, it reinforced a nagging insecurity: Was I good enough to compete in a man's world? Would I ever be good enough?

Fortuitously, I received an outside offer to join a botanical garden in Florida specializing in tropical canopy plants, especially epiphytes. I quickly did some homework on the quality of schools in Sarasota. There would be no move if I could not place the boys into an excellent public school, which had been a priority for our relocation to America in the first place. On a national scale, Florida had a bad reputation for K–12 education, but Sarasota had a public magnet school for math and science for which Eddie was old enough to qualify. The rest is history. We packed a small truckload of possessions, acquired some additional furniture from barn sales in upstate New York, and headed south, escaping the looming threat of navigating a difficult boss, especially tough in the tiny fishpond environment of a small college campus. Meanwhile, my divorce settlement came through and provided enough for a deposit on a modest house in Sarasota. (A conventional Australian divorce settlement was calculated at $11,000 per year for housewifery duties . . . below minimum wage!)

At the botanical gardens, I was the only female scientist—actually, the only female employee except one secretary!—in the research division. I was also the boss. It was not easy to win over all those "epiphyte guys," as I secretly called them, but it was necessary. The existing staff was composed entirely of taxonomists, and I sensed an undercurrent that my profession of ecology, which involved studying and conserving the habitat of epiphytes, was considered less worthy than their pursuit to classify them. So I jump-started several global programs to gain their respect—continuing research in Australia, but also executing new projects in the African and South American tropics. Despite its noble mission of studying and identifying tropical plants, the botanical gardens suffered from fiscal mismanagement and lots of deferred maintenance. I immediately ramped up its international reputation by hosting two global canopy conferences. They were both successful, attracting scientists from twenty-five and thirty-five countries, respectively. The city of Sarasota, historically associated with arts and culture, rejoiced at the renaissance of a "science garden," and the terms "epiphytes" and "canopy" became household words throughout southwest Florida. Soon after the second international conference, the board promoted me to the role of director. I went from their Indiana Jones of the treetops to sitting in their padded chair, and it was exciting to build community support for plants and forest conservation. I also felt a strong sense of duty to serve as a role model for girls; there were so few women in science leadership roles at the time, and I welcomed a chance to put a crack in the glass "canopy." I had already given a lot of blood to breaking glass ceilings, mostly in hopes that all those cuts might help the next generation of women colleagues avoid the bloodletting. Like a plant that is watered, I flourished with the job, and learned about staff management and budgets by taking an executive management course at Dartmouth's Tuck School of Business. I loved bringing plants to the public, and relished helping the staff achieve success, ranging from construction of a gorgeous wedding venue to planning a children's garden to updating development software to track donors.

After three successful years of fundraising, construction, grants, and expansion of the garden's footprint locally and globally, I had another encounter with the tall poppy syndrome. A new board chair came onto the scene, with a pedigree in the orchid-growing world. Given their volunteer status, it's understandable that board members looked favorably upon anyone with experience that appeared to align with the garden's mission. When a stunning purple orchid was imported from Peru and brought to the garden as a new species, this board chair seemed thrilled about the discovery. I initially shared some of the enthusiasm of the orchid staff, until it turned out that the orchid had been illegally imported.

During the week of the orchid's arrival at the garden, I was in

Ohio at a crew regatta in my other role, as mom. This did not help my leadership status, to say the least, because I still felt scrutiny as a female CEO, where exposing my motherhood duties was not perceived as professional.

The orchid taxonomists were so overly eager to acquire and describe the new species, they seemed to overlook the absence of any legal paperwork (or did they?). And the collections manager claimed to have lost its permit (or did he?), and then later professed to have never seen a permit, although in my view that responsibility was integral to his job. One of the orchid scientists was so enthralled by the species that I found out much later that he secretly took a piece of it to his home in Vermont to see if he could cultivate it. Even more audacious, there were reports of a Peruvian nursery owner appearing in the Miami orchid show not long after the new Peruvian orchid's arrival at the garden, allegedly offering the same species for sale at enormous prices. An international agreement called CITES (Convention on International Trade in Endangered Species) protects endangered plants and animals from being exported from their country of origin. Because the orchid staff had accepted a Peruvian orchid into their collection without a permit, they had essentially broke that law. At my urging, the garden hired a legal expert in Washington, DC, to sort through the complexities of the ensuing charges for accepting a nonpermitted orchid into the Orchid Identification Center, and the costs mounted. The orchid staff all pointed fingers at one another, and in the end, the board chair pointed a finger at me, even though I had been a thousand miles away at a crew regatta when the incident occurred and never did lay eyes on the specimen (even to this day). If the board chair's goal was to keep the orchid staff out of the spotlight, I guess the only woman at the table made a good scapegoat. While the board chair criticized my skills as a manager and others complained about the legal bills from the law firm I had brought in, I tried to hold some ground against this bullying. Several leaders in the Sarasota community sided with me, even fundraising to pay some of my personal legal fees. Refusing to fight a battle for something I did not do, I eventually resigned. One of the garden's orchid scientists wound up pleading guilty, as did the garden and, later, the man who had illegally imported the orchid.

It was a tumultuous episode and a stark example of how conservation can suffer setbacks from the very scientists working on the front lines to save species. In the case of these orchidophiles, their passion exceeded the boundaries of the law. But it also happens with other biodiversity. A similar incident arose in 2019, when Malaysian tarantula specimens were shipped to Britain and written up for scientific classification by some Oxford University affiliates. Coveted by spider taxonomists, it was a gorgeous cobalt blue tarantula. One of the arachnologists said that she and her coauthor were told that the specimens had been legally collected and the necessary documents existed, but they didn't. Similar to the Peruvian orchid, the tarantula specimens had been illegally imported and technically belonged to their country of origin.

The orchid experience hardened me to the world of board dynamics, institutional bullies, and snake-oil fabrications. It was hard to believe one beautiful orchid could inspire so much buck-passing among a group of otherwise dedicated botanists. As I looked around the boardroom for a sympathetic female role model, I saw none, and that sense of aloneness reinforced a conviction to always support fellow women in the workplace when hurdles confront them. It is not to say I have not had wonderful male role models throughout my career and am incredibly grateful. (Thank you to Peter Raven, John Replogle, Bob Ballard, Greg Farrington, E. O. Wilson, Hal Heatwole, Tom Lovejoy, and Brian Rosborough, to name but a few!) On the bright side, the experience as a botanical garden CEO who navigated a crisis opened doors to leadership positions at other museums and also reinforced my determination to work doggedly toward equity in science.

For eight joyful years, I served as a tenured professor of environmental studies at New College, Florida's honors undergraduate institution. But because of my past experiences communicating science to public audiences, I was recruited as the inaugural director of a new North Carolina museum wing and left my tenured professorship to return to the museum world. I also rejoiced in having my first female boss. Betsy was a force of positive energy, and the two of us probably each put in many hundred-hour weeks to fund, construct, and staff an innovative new museum wing. There was a tight timeline of two years before the opening ceremony, and I literally sprinted across the North Carolina landscape to follow her lead. One of the best successes was creating partnerships with different state universities for each of the newly hired science laboratory directors. These unique collaborations fostered vibrant student activity at the museum, as well as important academic status for the new curators. To meet the aggressive timeline, I organized a cluster hire, advertising and interviewing several curatorial positions at once. To get the best possible staff, I wrote a lively, compelling ad to indicate that this was not a conventional museum team. As a result, hundreds applied. Of ten successful new science hires, over half were women, a ratio almost unheard of in the museum world, where senior curators were still predominantly male. Although I did not prioritize either gender, I think more qualified women (than men) applied because, like me, they really wanted to work for a female boss. Another innovative achievement was the construction and programming of five labs with glass walls, so the public could directly observe scientists and their process of discovery. In addition, each museum scientist made frequent public science presentations in a new, cutting-edge theater, called the Daily Planet, shaped like planet Earth, and connected virtually to public schools throughout the state. (A year later, my Harvard colleague E. O. Wilson and I hosted a live conversation with all the middle schools in North Carolina from the Daily Planet.) With positive leadership, we not only finished the museum construction and its fundraising, but also hired a dynamic team just in time for a twenty-four-hour opening gala hosted by the governor, featuring an international cadre of dignitaries plus a parade along the main streets of Raleigh.

Just after the new museum wing opened, our beloved female director elected to retire, having fulfilled her life's dream and leaving it in good hands (or so she thought). I was crestfallen to lose my first female boss but crossed fingers the search committee would find a deserving replacement. Alas, the recruiting firm executed a secretive search with no staff input, at least as far as I know. Soon after the new director's arrival, I endured another hurtful episode of the tall poppy syndrome. Seeing the enormous momentum and success with donors and scientists alike, the new director apparently considered me a threat and conveyed to me behind a closed door that there was room for only one of us, and he was the boss. The next day, he thoughtlessly made a surprise announcement to the new staff about my demotion. He told them I would no longer supervise anyone or any budget, despite extraordinary success with both, and he was transitioning me into an "ambassador" for the museum with no official responsibilities. (As we all know in the business world, folks without a job description are the first to go in a budget cut.) When one recently appointed curator stood up at this meeting and respectfully explained they had all accepted their jobs in part for the opportunity to work for me, the new director cynically replied, "Well, now you can brag to all your friends you work under me instead." A few of the staff cried, and others became angry. During this same meeting, several other staff members made eloquent pleas to his deaf ears. It was a very tough time for a newly minted staff at an exemplary state museum that had just achieved the highest accolade as North Carolina's most visited destination, plus a presidential medal. It was overwhelmingly evident that hard work had earned me the respect of the North Carolina public as well as the academic community, from TEDx Talks to a major collaborative National Science Foundation grant. Had there existed a strong sisterhood of other executive women on my speed dial, could I have strategically redressed this new director's workplace insensitivities? I will never know. My passive response was simply to look for work elsewhere, believing it was in the best interest of an institution I loved. On an economic note, I was only six months short of qualifying for a state pension. Looking back in the files, I found a commentary from an online publication called *World of Geology and Earth Science*, written in 2013 by Professor Anne Jefferson of Kent State University, whom I have never met but who has graciously given permission to reprint this. Her words summed up my tall poppy episode perfectly:

As my daughter plays paleontologist in the next room, I am thinking about three stories from the last few months. They are stories that illustrate why, despite the progress women and minorities have made in the past few decades, we still have so far to go before we get anywhere near true gender parity in science and leadership. They are stories that show how far some people will go to silence women and minority voices, and how those silencers are in positions of power or aided and abetted by those who are.

In June, rock star scientist "Canopy" Meg Lowman was, with no public explanation, stripped of her directorship of the Nature Research Center at the North Carolina Museum of Natural Sciences. As director of the Nature Research Center, Lowman, an incredibly distinguished scientist in her own right, was the supervisor of scientists working in innovative ways, directly with the public. The mission of the NRC is to "bring research scientists and their work into the public eye, help demystify what can be an intimidating field of study, better prepare science educators and students, and inspire a new generation of young scientists." What better person to help the center succeed at its mission than its charismatic director, Dr. Lowman? But instead, the museum's leader demoted her to Senior Scientist, took away her direct supervisorial responsibilities, and spent a bunch of ink emphasizing how she would still be a "female leader" and a "role model to girls and women in science." Personally, as a woman in science and mother of a girl in science, I thought Dr. Lowman was much better able to be a leader and role model as director of the research center than just another senior scientist. But clearly, those with power didn't want her voice and authority to be too loud. It's OK for Dr. Lowman to talk about "women stuff" but she shouldn't presume to go beyond that.

In a rescue mission, one of the new female curators nominated me for a senior position at another museum, fervently hoping I could continue in a position of influence for diversity in science and creative leadership. I moved to California. Like North Carolina, I accepted this new position in part due to another visionary, charismatic boss, but was again disappointed. Believing he had put the perfect leadership team into place with me as his last senior hire, this museum director announced his retirement soon after my arrival. Once again, the search team tasked with finding his replacement was secretive, and no one knew their choice until he walked through the door. (Sound familiar?) The new director was not only new to the museum world, collections, and fundraising, but his credentials mirrored mine as an ecologist and science communicator, with an almost identical Rolodex of colleagues. But he was not a muddy-boots scientist, as were the museum curators. I suddenly became a glaring tall poppy in his new world, and even worse, was actually more experienced in museum administration and collections. His leadership style was temperamental and top-down. He publically denigrated our CFO, so she left. Similar episodes happened with a few others. Then he turned to me, and I could feel the icy vengeance. Within a two-year period, I was demoted and replaced by a young neophyte who had almost no leadership experience and lacked the years of scientific networking required to bolster the institutional brand. Even worse, I was given three different job titles and salary cuts in rapid succession, strategically eliminating much hope of success. I threw up after work and suffered panic attacks in the middle of the night. Despite pay cuts and title changes (some so rapid that Human Resources never even wrote up a new job description), I remained highly productive in terms of publications, funding, and global collaborations. Other museum colleagues were aghast and quietly hugged me, especially the females I had recently hired, now lacking a senior mentor.

When toxic leadership infuses an institution, it often taints the air for a long time. Issues such as staff anxiety, disenfranchised members, unsuccessful fundraising, or loss of morale can take years to recover from. And worse, such workplace toxicity sometimes persists in a handful of staff who learned how to manage or even manipulate the toxicity. At the height of the museum's leadership turmoil, I was targeted not just by the director, but also by his young protégée. When I was her supervisor, I had generously supported her through compensation as well as hiring her husband, to ensure she had a supportive homelife. Her reverse treatment of me felt worse than a machete attack. It was heartbreaking for the new female hires to witness a woman unwilling to support another woman, and certainly tough to be the scapegoat. When she handed me a list of petty complaints, I decided not to engage in nitpicking to correct false accusations. Behind closed doors, she told me my research involving conservation and sustainability were no longer her institutional goals. So I made the tough decision that I simply could not give up an enormous portfolio of research based on my two strengths, and resigned. As bittersweet irony, the museum's annual report six months later focused on one of my major conservation projects of funding a team of museum curators to survey biodiversity of a Malaysian tropical rain forest, culminating in its UNESCO World Heritage nomination.

After fifty-two years of building up a career from a few wildflowers in fifth grade to achieving international recognition, it was heartbreaking to be clipped like a tall poppy. In a 2018 *New York Times* article reviewing the National Science Foundation harassment policies newly issued by its director France Córdova, gender harassment was defined as "verbal and nonverbal behaviors that convey hostility, objectification, exclusion or second-class status," and the article further explained that this form of harassment is far more common in science settings than sexual coercion. A 2020 survey by the Wellcome Trust further reported that of 4,200 scientists, 60 percent were bullied by supervisors, and over 40 percent by colleagues. Having confronted this behavior, I only wished I had adopted the traits of a strangler, not just a benevolent fig. By not standing up to bullies, I enabled them, and allowed myself to become a victim. The #MeToo movement was a welcome advance for women in all professions, but it came too late for some of us. The absence of females at conservation decision-making tables has also been identified as a setback to global problem-solving, and such a shortfall probably mirrors the scarcity of women in leadership roles. *Nature* magazine featured an article in 2016 by Heather Tallis and 167 others, me included. It explains that women are underrepresented in conservation leadership. The authors suggest such inclusivity will ensure more progress toward achieving effective conservation, but progress is slow. The World Economic Forum's *Global Gender Gap Report* in 2016 calculated that, at the current rate, 170 years will be required to achieve equal salaries between genders worldwide. (I posted this article on the wall behind our museum photocopy machine, and it was removed, not once but twice.) The same report documented that during the second decade of the twenty-first century, the United States failed to rank in the top twenty-five countries for improving pay disparities, coming in at forty-fifth place. And to confirm industrialization does not automatically equate to gender balance on the payroll, Rwanda was in the top ten with progress for women's economic equality, a country that also boasts 64 percent female parliamentarians (the highest in the world).

With quiet dignity, I packed up my blowguns, volumes of data, vast Rolodex of international scientists, and relocated back to Florida to continue global conservation as a freelance explorer-author. The mangroves, the year-round sunlight, and the shorebirds provided salves for all those cuts and bruises incurred from a few glass canopies. As a firm believer that the cup is always half full, not half empty, I felt confident that change would undoubtedly bring new opportunities. One such blessing was the chance to be near my aging parents, who'd relocated to Florida, and share quality time during my dad's last year of life and reciprocate the steadfast support my mom had so generously offered me throughout a lifetime. I've named this new career chapter "5,000 Days," based on a calculation of the approximate duration of my remaining productive work life. I want to make each day count and not lapse into political miasma or meaningless paper-pushing. As Booker T. Washington so wisely said, "Success is to be measured not so much by the position that one has reached in life as by the obstacles which [they have] overcome." The world needs bold leadership to put the planet on track for our children and grandchildren. I have resumed pro bono directorship of a conservation foundation where every dollar is spent saving forests, not on office real estate or overhead. And this book is a major focus of those five thousand days; I wanted to share my story of a life spent in research, discovery, and exploring the wonders of trees such as figs. Through their incredible fruits and spreading crowns, figs nurture an entire ecosystem. Women (and men) must follow their lead and make the best use of resources, engage in productive outcomes and interactions, and nourish their communities.

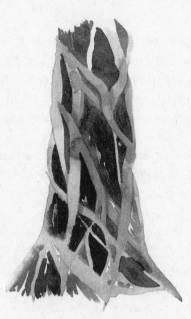

⇶ Figs ⇷
(*Ficus* spp.)

ONE OF MY FAVORITE WALKWAYS in the world spans a beautiful fig in the village of Falealupo on the island of Savai'i, Samoa, formerly Western Samoa. My colleague the ethnobotanist Paul Cox invited me to visit this island in 1994 at a critical juncture in its history. The sixteen tribal chiefs were at a crossroads. The Western Samoan government required villages to build cement schools because their former palm-thatch construction could not withstand the frequent monsoons. But the cost of a school was over $50,000, and Savai'i did not have a cash economy. Samoans are incredibly supportive of their children and wanted the best possible education for them, but their livelihood was based on harvesting fish from the sea and fruits from the jungle. An Asian logging company offered to harvest the island's timber in exchange for enough funds to build a new school. The chiefs were uneasy because their entire existence over many generations relied on the forest, and even their ancestors were part of this ecosystem, returning to earth as flying foxes in the canopy. So I flew to Samoa with two construction engineers to discuss the notion of a walkway. I was honored to be included in a ceremony with all sixteen tribal chiefs, where they sang, drank their sacred kava, and discussed the options of logging versus ecotourism with regard to the fate of their island trees. After more than five hours, they unanimously agreed to take out a loan and build this newfangled idea called a treetop walk. They had never heard of ecotourism before but decided to trust the concept our team had proposed: that visitors would pay to explore the island's eighth continent. Today, a platform circles the island's largest fig, connecting to a bridge that spans to the local school. Within two years, the debt for the new school was repaid by ecotourism, not logging. As a thank-you, the senior chief anointed me "Mati," which means "fig," and presented me with a sacred kava stick, now a favorite treasure. I have always admired and loved figs, but in this case, one emergent Samoan fig literally saved the trees of an entire island.

Theophrastus, an ancient Greek philosopher who studied plants (circa 300 B.C.), is credited as the founder of botany, and also the first person to describe fig species. He focused on *Ficus carica*, an important edible fig, and was unaware that hundreds of others existed, each with its own life story. He did learn about one particularly important banyan (*Ficus benghalensis*) from the stories of one of his contemporaries, the conqueror Alexander the Great, who claimed that ten thousand soldiers sheltered under one fig, with all its pillar roots forming an enormous leafy umbrella. Much later, in the early twentieth century, the English botanist E.J.H. Corner specialized in *Ficus*, and described hundreds of them. One of Corner's unique legacies was training four monkeys to fetch fig fruits in the canopy, so he could describe new species without climbing himself. In a sense, these monkeys represented an early arbornaut's toolkit, and he affectionately called them his "botanical monkeys." Within six months, they collected the fruits of 350 species, far more than a human climber could.

A member of the diverse mulberry family (Moraceae), the Latin genus *Ficus* includes approximately 800 species of deciduous or ever-green trees, some of which are cauliflorous (meaning they flower and fruit along their main trunk instead of at the ends of branches), plus a few shrubs and vines. An estimated 150 figs inhabit the New World (Americas), and over 600 species in the Old World tropics (India, Asia, and Australia). Most are edible and germinate in the conventional way, starting life as a seedling and growing upward. Any fig that begins life as an epiphyte is called a banyan, technically part of the subgenus *Urostigma*, and includes the well-known strangler figs. In my global exploits as a field botanist, I found that stranglers occupy the most extraordinary lifestyle of any trees on the planet, bar none! The name itself conjures up the antics of James Bond or some Sherlock Holmes murder mystery, but the actual life history of this species engenders shock and awe. During his exploration, Alfred Russel Wallace called stranglers "the most extraordinary trees of the forest."

My first exposure to the incredible strangler figs came when I climbed them in Australia, most notably the iconic emergent *Ficus watkinsiana*, named for George Watkins, the former president of the Pharmaceutical Society of Queensland and plant collector. He was an Englishman who came to Queensland in the 1800s and trained as a pharmacist. He pursued natural history in his spare time and participated in numerous expeditions. In 1891, F. M. Bailey named a fig species after him. It was called Watkins' fig, nipple fig (due to the fruit structure), or green-leaved Moreton Bay fig. Its fruits are rounded, purple-black, and have distinct "nipple" structures, although the word "fruit" is deceptive because figs are anatomically not really fruits but a cluster of hundreds of flowers enclosed in a smooth skin. Like other fig species, the trees have male and female flowers on one plant, known as monoecious, from the Greek word for "one house." *F. watkinsiana* leaves are tough, waxy, and entire, although a few fig species have lobes. Another famous Australian banyan is the curtain fig outside Atherton, Queensland, which occupies close to an acre of ground with all its aerial roots, though it's not nearly as vast as Alexander the Great's historical tree.

Different from most tree seeds that germinate on the forest floor, which is relatively dark and inhospitable to a delicate seedling, stranglers start life at the top with plenty of sunlight. When a figbird ingests fig fruits, it defecates the seeds onto an upper branch. This unique beginning means stranglers literally start life as an epiphyte with adequate light, eventually sending down roots to absorb water and nutrients below. The limiting factor for a strangler seed tends to be moisture, so germination is most successful on rotting or decaying bits of branches. My fellow arbornaut Tim Laman of National Geographic studied the germination of one strangler fig (*Ficus stupenda*) in Asia. He placed its fruits in the canopy to determine ideal conditions for downward growth and found they grew best in moist, decaying branches or wet bark crevices. Tim also documented the many animals that disperse figs, including hornbills, gibbons, orangutans, and flying foxes.

Once a strangler is established in the crown, a latticework of downward-growing stems thickens with time and often surrounds the host, sometimes creating a hollow center if the host has soft wood and subsequently rots inside. In some cases, strong-wooded host species survive and live in tandem with the strangler, so both trees are intertwined as a massive duo. Although stranglers, or banyans, often constrict their host, they also sustain thousands of other creatures in the ecosystem. It seems astounding that other species have not evolved this adaptation to outcompete for light, space, and water by starting at the top. Seedlings that germinate on the forest floor must wait patiently, sometimes decades, for a treefall to create a light gap, plus have the good fortune of not being trampled, eaten, or buried during their precarious juvenile phase. Some figs eventually form buttresses, which are aboveground prop roots that stabilize their shallow-rooted shapes; they would otherwise be prone to fall during monsoons.

Mature *Ficus* (including banyans) produce thousands of fruits, an important food source in tropical food webs. Fertile figs, like the ones we eat, are distinct because of their unique fleshy, hollow receptacle with an inner surface laced with tiny flowers or fruitlets. All species have tiny male and female flowers that develop on the inside wall of a syconium, which eventually becomes a fruit. A tiny female wasp of the Agaonidae family enters a hole in the top of the syconium and passes through a tunnel lined with downward-facing scales that only allow one-way travel. The wasp finds two types of female flowers inside: stalked, short-styled flowers and long-styled flowers that are not stalked. While she pollinates the flowers, the wasp also lays her eggs and then dies. After a few weeks, her offspring hatch, and wingless males mate with winged females, then chew an exit tunnel that allows the females to fly to another fig and repeat the process. The females pick up pollen from recently matured male flowers as they exit. Some botanists originally thought all 800 species required a specific pollinator wasp, but more recent molecular analyses reveal that 119 fig species shared multiple pollinator wasp species. When a wasp dies inside the fig after pollination, enzymes (called ficain) break down the carcass; consequently, anyone who eats figs is also consuming wasp biomass. When scientists at the National University of Singapore experimentally exposed fig wasps to warmer temperatures, their lifespan significantly shortened. This suggests climate change could jeopardize *Ficus* pollination success, if wasps don't have enough time to find their specific fruits. As always, it is difficult to extrapolate from the laboratory to the forest.

As with most complex interactions, additional players in the fig-wasp interactions are parasitic wasps, some of which also lay eggs inside figs, which prevents the beneficial pollinator wasps from entering. Even worse, some parasitic wasps inject their eggs through the fig's surface without even boring a pathway inside, and their offspring feed on the pollinator wasp larvae. Hundreds of wasps function as pollinators, and each *Ficus* species is known to host up to thirty-five species of non-pollinating wasps, making their crowns a hot spot of biodiversity. In addition to wasps, over twelve hundred recorded species of birds, insects, bats, and mammals, including humans, eat the fruits, making *Ficus* a true keystone species for the biodiversity of many tropical forests. In the Amazon, ornithologist John Terborgh wrote that if figs disappeared, the entire ecosystem could collapse. Biologist Dan Kissling correlated the number of fig species in sub-Saharan Africa with the abundance of fruit-eating bird species, and concluded figs were the keystone resource.

Figs have many uses for humans as well as for insects, birds, and animals. Their milky latex contains medicinal properties, and in the Amazon, villagers administer a teaspoon to cure stomach parasites in children. According to indigenous people in French Guiana, Colombia, and Brazil, the same sticky latex can cure cuts, fractures, and abscesses. In Nepal, *Ficus benghalensis* leaves, bark, and roots are used to treat more than twenty ailments; this useful banyan is also used in India to treat a wide variety of disorders ranging from tooth decay to hemorrhoids, diabetes, and constipation. At one time, figs were grown for rubber latex in India and parts of Africa. Probably the most famous fig is *F. religiosa*, or the Bo tree, native to India and Southeast Asia, thought to be the canopy under which Buddha received enlightenment. This is one of several species that in India and Asia dominates the center of many rural villages as an important spiritual gathering site.

Embarking on a new role as grandmother this decade, I have decided to use my Samoan nickname of Mati instead of Grandma; I will tell my grandchildren about all the wonderful figs I have encountered. Maybe someday I'll whisk them off to the island of Savai'i in Samoa to see an incredibly special walkway in an emergent fig or to climb a few favorite stranglers in Australia.

7

ARBORNAUTS FOR A WEEK
❧
Citizen Scientists Explore the Amazon Jungles

I T WAS DARK. PITCH-BLACK. Our small dinghy, called a peke-peke
by the locals, held twelve people despite its tiny propeller engine,
and it churned ruggedly through muddy water swirling with flot-
sam of branches and logs while my citizen science team engaged all
five senses to absorb every element of this forest. Many of them had
never experienced absolute darkness before; there were no lights for
miles, except the odd phosphorescent fungi twinkling like reverse
constellations on the jungle floor. No urban noises of sirens or traffic, just a surround-sound symphony of frogs, katydids, crickets, and
an occasional piercing note from the tinamou, an elusive nocturnal
bird. No toxic odors of human pollution, just the heavy aroma of
earthy, rich-chocolate humus and the intoxicating perfume of a thousand trees flowering at once. We were traveling through the Pacaya-
Samiria National Reserve, a sanctuary of over five million acres above
Iquitos, Peru. Gliding slowly along a tributary of the majestic Amazon
River, we paused occasionally and turned on our headlamps to search
for caiman. Measuring up to fifteen feet in length, these camouflaged
alligator relatives swim at night, eyes glowing bright red in the dark
and poking just above the waterline. A few spider eyes also caught

the glint of our lights as those hairy, eight-legged predators dangled over the water, awaiting dinner to jettison into their enormous orb webs. I casually asked our local guide and my longtime friend Guillermo if we might spot an anaconda during the trip. He paused, gazed out along the dark riverbanks, and suddenly leaped overboard. After a bit of splashing around, he emerged and clambered back onto the boat holding a juvenile anaconda in his bare hands! We all gasped, cautiously eyeing this youngster of fifteen feet. Our guide had seen the telltale bubble of air released by a submerged anaconda and jumped on top of it, quickly grabbing its head and tail, even in darkness. We were in awe not only of this magnificent creature, but also of Guillermo's thorough knowledge of local wildlife—the kind that can't be taught from a textbook or in a college lecture hall. After everyone snapped photos, the anaconda was gently returned to its murky aquatic home. I count my blessings to have friends who grew up along the world's mightiest river. In the Amazon, life depends upon knowing how to hunt and where to find essential natural resources, knowledge that is now transforming conservation through its application to citizen science and education.

This trip reminded me of one of the major reasons why I became a scientist in the first place. Part of the joy of scientific discovery is sharing it with others. And linking people to nature is one of the most important components in addressing conservation challenges. Using multiple pairs of eyes and ears in field biology can result in more data points and greater breadth of information, plus give participants personal investment in the work of science, making them better ambassadors for Earth. One of my favorite places to engage the public is in the world's largest canopy, the Peruvian Amazon jungle, a place where you can still find so much biodiversity, in descending order of relative abundance almost like the twelve days of Christmas: at least five trillion leaves, one hundred billion ants, one million beetles, one thousand orchids, three hundred bats, one hundred spider monkeys, sixty-three delicious piranhas, thirty-one enthusiastic explorers, ten pink-toed tarantulas, two anacondas, and one emergent green bean tree with its own aerial platform.

Citizen science expeditions to the Amazon increasingly center around one of the world's longest skywalks, called the Amazon Conservatory for Tropical Studies (ACTS) Canopy Walkway. It spans almost one-quarter of a mile with twelve bridges and thirteen platforms that extend to a height of 125 feet. Called the eighth architectural wonder of the world by forest scientists, this treetop trail provides otherwise impossible views into the lives of millions of otherwise invisible species. Like many others built since, our first walkway back in Australia used poles to support its bridges spanning away from a slope to a height of fifty feet. But in the Amazon, no poles were tall enough, so one of two techniques was required for suspending bridges: (1) through-bolt installed through the trunk like a piercing, to hold stainless steel cables in place but with no harm to the tree; or (2) necklace of cable encircling the girth, with rubber pads to buffer the bark from damage. A through-bolt is safer for tree health, because most of the trunk is dead wood and not impacted by the hardware. A tree's only living cells are a thin layer of vascular tissue that sits just below the bark, so a circular cable can easily damage those vital cells by strangling the trunk. Given the enormity of trees and the remoteness of base camp in the Peruvian Amazon, it was not possible to haul large generators to these sites, nor could we find long enough through-bolts. So loops of cable were gently affixed around twelve strong trunks and buoyed away from the vascular tissue by rubber grommets to minimize strangulation. It took six months to build the first platform of this walkway because the locals only had manual drills. But after a simple battery-operated drill was brought to the site, the next ten platforms were completed in just three months. This timeline is vastly different from similar construction in forests closer to big cities, where the building process of an aerial structure may take only a few weeks yet the permitting can take a year due to liability and paperwork. A creative Maryland engineer, Ilar Muul, designed the Amazon walkway using the necklace technique. This structure requires a lot of careful maintenance to constantly check for termites and fungus under all the rubber grommets encircling the trunks. Fortunately, the local villagers, proud of their ecotourism operation, inspect the structure almost daily as part of their employment as guides.

In the Amazon jungle, this extraordinary skywalk allows citizen science teams to experience the whole forest. Travelers have included students, educators, community leaders, CEOs, families, and scientists aged seven to ninety. In the eighth continent, everyone is an explorer! As a young field biologist in Australia, I had the good fortune to be indoctrinated to citizen science through Earthwatch, and after that engaged the public in many science projects as a museum leader. Perhaps my favorite citizen science program involved middle school kids and was called the JASON Project. After he found the *Titanic* at the bottom of the Atlantic Ocean, marine explorer Bob Ballard received thousands of letters from schoolchildren, all inquiring if they could accompany him on his next trip in the *Alvin* submarine (which only seats two!). Bob realized he could never directly transport thousands of young people undersea for marine research. Instead, he founded the JASON Project, named after his favorite marine hero, Jason, of Golden Fleece notoriety. Bob used satellite telecommunication to link remote field expeditions to schools and museums. When he needed a terrestrial scientist to balance his marine expertise, Bob tapped me for the role. I earned the nickname "Canopy Meg" from Bob and the millions of kids who participated in the JASON Project. Bob and I both marveled at our different worlds—my love of canopies despite the insects, perspiration, humidity, and venomous creatures versus his passion for the undersea world despite the claustrophobia in a submarine plus challenges of darkness and pressurization. Bob was a great role model for enthusiastic science communication, even if he did occasionally swat some of the insects I was trying to measure. And soon enough, we both realized this was great training for us too: if you could speak clearly to seventh graders, you could also communicate effectively to politicians, who often had about the same level of science training. We broadcast virtual expeditions in the tropical forests of Panama and Belize, where I oversaw construction of a new walkway for the broadcast and our team figuratively followed a raindrop from the canopy to the forest floor and out to a coral reef. In Panama, I partnered with a soil scientist, affectionately calling our broadcast "The Scoop on Poop" because we measured insects eating leaves and followed their frass down to the soil as part of nutrient cycling. Those early JASON expeditions cost millions of dollars; we had to ship a satellite dish via oceanic barge, transport thousands of pounds of video cameras and electronic wire to string through the forest, and hire over twenty technicians to operate a production studio. (Today, we broadcast remotely at less than a tenth of the cost using a single laptop with internet connectivity and two or three technicians.) The JASON Project hosted its 1999 broadcast season from the Amazon walkway. After fifty-three hour-long broadcasts to over three million students from a 125-foot-high platform, I was not only sunburned and bitten by bugs but also transformed into a STEM celebrity in the eyes of many middle school science teachers. Following the broadcasts, a few wrote asking if they could assist with my research. Amazon citizen science expeditions soon became an annual event. For more than twenty-five years, teachers, students, families, and CEOs have joined me in the Amazon to conduct long-term field research. Those volunteers have calculated herbivory, discovered insects, measured leaves of different species and heights, and estimated biodiversity. On these citizen science expeditions, everyone contributes to our collective knowledge of tropical jungles, which translates via real and varied ways into conservation.

The walkway itself inspires botany lessons about the majestic trees that support its twelve platforms. Each species, plus all their inhabitants, tells a story and has a function in this complex tropical forest canopy. For example, strong trunks of *Inga* sp. (Peruvian name shimbillo) hold up several aerial platforms and also provide timber for local people to construct huts or carve canoes. Their compound leaves have a small cuplike gland between each pair of leaflets, whose function remains unknown and a feature I never saw on any other plant. Legend has it that butterflies drink from these tiny receptacles, and my students fantasized it could also be a drinking cup for fairies. Over 350 species of the genus *Inga* exist throughout the Amazon, yet scientists still do not know much about its ecology, except that citizen scientists carefully calculated that its foliage was delicious to herbivores, with up to 30 percent surface area consumed. *Inga* grows best in moist lowland forests, but with the onset of climate change and the warming of the Amazon, scientists have observed them dying more rapidly than other, more drought-tolerant species. Another common platform support tree is *Apeiba membranacea* (Peruvian name *peine de mono*). *Apeiba* fruits resemble sea urchins and are coveted for use in local crafts. Its bark has a pH conducive for bromeliads and orchids to populate its branches, creating a vertical festoon of colorful greens and pink flowers.

A third species that carries its own mystery is *Cedrelinga cateniformis* (Peruvian name tornillo), which grows to over 150 feet high. I affectionately nicknamed this species the "green bean" tree, to remind students it is a close cousin of their dinner vegetable and a member of the legume family, which ranges from small six-inch-high vegetables to large two-hundred-foot giants. This emergent supports the walkway's highest platform at 125 feet, high above the forest floor, which was an aerial base camp for the JASON Project. Each day for two weeks, a movie crew filmed me measuring herbivory (and swatting sweat bees that landed all over my face). The research was livestreamed to middle school students around the world while host Bob remained on the forest floor, asking questions all the while that were relayed via microphone. A family of shovel-tail lizards (*Tropidurus flaviceps*) lived on the branch next to my roughshod research table, and I got to know both parents and all five offspring in their aerial territory, where they feasted on insects that had feasted on me. (Descendents of that same lizard family are still there twenty-five years later.) One of the most often-asked questions by K–5 students during JASON broadcasts was, How do you manage to go to the bathroom in the canopy? Answer: a hasty trek across six bridges and five platforms to descend the stairwell to the forest floor. For girls, there were no shortcuts, although the film crew (all male) each had a jug used discreetly in situ. Relieving oneself from such a height is simply not polite, especially when there's a well-traversed ground trail underneath.

Another unique attribute of JASON was that while millions of middle school students participated virtually, several dozen assisted the scientists during the broadcast. They were recruited from local as well as international schools. In this emergent roost, I hosted a fifth grader from Iquitos, Peru, who had never before visited her local tropical jungle. Pamela became an expert at measuring leaves and swatting sweat bees at the same time. Fifteen years later, she earned a scholarship to the University of Florida to study environmental education and ecotourism, inspired by her JASON experience. Another student achieved fame by winning our online competition to name a new species of beetle discovered during the broadcast. A panel of scientists voted on the one thousand entries submitted and selected hers: nutmeg beetle, named for the host tree (in the nutmeg family), the beetle's color (nutmeg), and its finder (Meg), all embedded in the name.

As an emergent, the green bean tree provided a lookout from which to see the rest of the forest. A variety of tropical flora were visible from our platform, some with wonderful names and amazing medicinal uses. *Oxandra xylopioides* boasts cauliflorous flowering habit and attracts mid-canopy pollinators; palms *Lepidocaryum tessmannii* and *Astrocaryum* sp. are both tongue-twisters and important plants for the shaman's jungle pharmacy. *Virola* sp. provides a home to Azteca ants, which give a walloping sting if you try to pluck a leaf; the elegant *Symphonia globulifera* with its cherrylike fruits is so colorful and distinct during flowering season that a drone could be deployed to count the local population of red crowns.

After my first summer of teacher workshops at the Amazon walkway, I devised several citizen science units so volunteers could contribute to tropical research in a meaningful way. This is tricky business. It is not useful to ask amateurs to carry out complicated actions, such as writing technical classifications of insects or finding critters that require a trained eye to detect. Instead, activities need to be simple enough that they won't contribute erroneous data to the overall research, but interesting and meaningful enough to provide a sense of reward to the participants. In this case, I invited volunteers into the canopy to find, photograph, and sometimes collect insect herbivores. Forty eyes could see more deeply into complex foliage than just my two eyes. Working in teams, citizen scientists collected leaves and learned how to measure herbivory by a simple process of counting graph-paper squares and in turn calculating the squares of proportional leaf area missing. Back at base camp, they played out this painstaking process in a place with no electricity—the almost-primitive counting methodology became a cult of sorts. Each team adopted a tree, collected thirty leaves, and calculated its defoliation. I called it the Leaf Lovers Club, which now boasts hundreds of international members. The data were entered into a spreadsheet on my laptop, so the comparisons between species provided information about which species are most resistant to insect attack. The local shaman and I enjoyed discussing these results—we both appreciated that plants produce chemicals to repel insects. Those species with little or no herbivory are usually the most toxic, which typically translates into their being important medicinal plants.

Our first several summers were teacher-exclusive workshops, but since then my citizen science Amazon trips have been open to a diverse public. With students and volunteers, I published articles about Amazonian epiphytes and herbivory, calculated the damage of leaf miners in the canopy for the first time in the annals of science, and assessed the local economics of the walkway in creating sustainable employment for indigenous people from ecotourism instead of logging. Every expedition starts in Lima and connects via local airlines to Iquitos, a river city in northern Peru approximately twenty-two hundred miles from the mouth of the Amazon and with no road connections to the outside world. The Amazon is called the river highway since it connects everyone and everything in northern Peru, from markets to medicines to marriages. From Iquitos, we boat for about five hours downriver and then up a tributary called the Napo (close to the border of Ecuador), watching for pink dolphins at every watery intersection. Our destination, the Amazon Conservatory for Tropical Studies, is an international field station situated on the edge of a million-acre preserve of primary tropical rain forest in northern Peru. At ACTS, botanists have recorded over 750 species in one square hectare (2.47 acres), almost a world record. A research camp houses up to forty visitors at a time, without electricity or running water but with the world's best wildlife symphonies and possibly the freshest air on the planet. ACTS partners with a local nonprofit called CONAPAC (Conservacion de la

Naturaleza Amazonica del Peru, A.C.) for activities including fresh-water filtration in villages, environmental education in schools, and sustainable incomes for local villagers through ecotourism.

Located at latitude 3° south, this section of the Amazon receives over two hundred inches of rain per year, and temperatures hover in the eighties with average humidity around 80 to 90 percent. According to surveys, biodiversity in Amazon forests exceeds that anywhere else on the planet and will be even higher when canopy inhabitants have been tallied. Approximately twenty thousand plant species grow in the United States, but over eighty thousand plant species live in the Amazon, along with over two million insect species, twenty-five hundred fish, and fifteen hundred birds. Why are the tropics so rich in life? Hypotheses include a mild climate, the complexity of three-dimensional niches in a tall tropical forest, and the long evolutionary timelines that have allowed radiation of many species in fairly stable environments. Living along the riverbanks, the local people, or ribe-reños, cultivate land, harvest food, and then move if the river floods or alters course, just as they have done for many generations. An Amer-indian tribe called the Yagua lives near our base camp and share their knowledge of the forest with my teams. Two of their most important plants are chambira palm (*Astrocaryum* sp.) to weave bags, hammocks, and decorative jewelry, and Irapay palm (*Lepidocaryum tessmannii*), which grows in the understory and is harvested sustainably for roof thatching. The locals utilized an estimated five hundred thousand palm fronds to make our dining room roof, which resists the incredi-bly heavy rains.

The definition of citizen science is any action where nonscientists assist scientists in conducting research. Volunteers have perspired, overheated, lost sleep, and eaten bugs, all in exchange for engaging in exploration and discovery. Even in their downtime, it's never a dull moment, from encountering tarantulas in the shower, to seeing ana-condas by night, to fishing for piranhas. In this biodiverse ecosystem, more questions abound than answers. How do so many millions of species live in one place? Why don't animals devour all the leaves, since trees cannot run away from hungry beetles, ants, and sloths?

How do tiny orchid bees navigate in a sea of green to find their specific flower? What tips off a shaman to discover the best medicinal plants? In the Amazon jungle, survival of the fittest is not an abstract concept. Both plants and animals evolve strategic behaviors and defense mechanisms to avoid being eaten, overshaded, trampled, strangled, dried up, outcompeted, or infected. As with travel to most remote places, discovery lurks behind every trunk. To survive in a tropical rain forest, camouflage is the name of the game.

One notable element of Amazon field research is the heat! The uppermost canopy is extremely hot and dry, with occasional heavy rain showers that pelt and sometimes tear the leathery leaves, but droplets quickly filter through to lower foliage and the forest floor. The tree crown is like an aerial desert; downpours rapidly turn into throughfall, then the top layer remains hot, dry, and windswept. One group of tropical epiphytes includes cacti, adapted to living in blazing sun without much water, conditions found in both the desert and a tropical rain forest canopy. It is so sultry and humid that sometimes our clothing needs wringing out after becoming drenched in sweat. I occasionally daydream about transforming into an arctic biologist, so my clothes would not always be growing mildew. The locals bathe in the river several times a day, and have no fear of piranhas, many of which are vegetarian and quite harmless despite their Hollywood portrayal as bloodthirsty predators. The showers at the field station are cold but refreshing, and feature river water pumped through a simple unfiltered pipe, minus the piranhas and anacondas. (But be careful not to open your mouth in the shower lest you risk getting sick from tropical microbes to which our bodies are not adapted!)

Leaves, those trillions of tiny green machines that form the basis of all life on Earth, overwhelm our five senses in all directions in the Amazon. We see greenery, we smell decay, we touch leaf hairs, we sometimes ingest medicinal plants. The citizen science mission for my Amazon expeditions, however, is to survey insect damage as a clue to gauge forest health. After twenty-five field years, my volunteers have confirmed that herbivores annually consume over a quarter of Amazon rain forest canopies, similar to findings in Australian rain forests.

Given the millions of insects in the crowns, I guess this is not really surprising, and anyone who ascends into the treetops returns with a huge admiration for "holey leaves." Upon close inspection, it is difficult to find even one leaf not chomped or sucked or tunneled. On average, only twenty-one leaves per one thousand are intact (i.e., without a bite). The Amazon foliage has recently been nicknamed a "smoking gun" in the mystery of how moisture and rainfall affects forests. For a long time, scientists could not explain why the onset of seasonal rains began a few months before the oceanic currents brought moist air from the oceans. The leafing-out period for most tropical forests contributes significant water vapor into the atmosphere as a result of so much photosynthesis. During transpiration, leaves release moisture from small pores called stomata, enough to create low-level clouds, detectable above the forest by NASA satellites. These leaf-induced rain clouds cause showers that in turn warm the air, triggering wind patterns bringing additional moisture from the oceans, linking rainfall cycles to the leafing patterns of tropical trees. Hooray for leaves!

Like Australian rain forests, most Amazon sun leaves are small, tough, and short-lived, whereas shade leaves are darker, larger, and more long-lived in the understory. Regardless of size, shape, or age, all foliage transpires to a certain degree and thus adds moisture to the atmosphere. But new foliage in direct sunlight makes the biggest contribution to moisture cycles. One crazy plant that breaks the conventional rule that leaf size decreases from understory to canopy, however, is philodendron (*Philodendron* spp.), the same plants that often decorate dentist offices. Its sun leaf is as large as a bath towel, whereas understory foliage is smaller and more like a washcloth in size. Smaller leaves are physiologically better adapted to survive the rigors of the hot, dry, and windy treetops, so how can enormous elephant-ear-sized leaves avoid drying out and over-transpiring? (Here is a PhD research question for an eager botany student!) Philodendron is also a hemi-epiphyte, spending the first half of its life as an epiphyte and the second half with roots affixed into the ground. In our research, its average sun leaf was a whopping 6,088 square centimeters (200 square feet), with only 1.8 percent herbivory. Its lofty, sunny location appears to protect it—the tissue usually remains relatively uneaten, and through some unique physiology, manages to avoid wilting or drying out completely.

After a guesstimate that trillions of leaves inhabit a tropical forest, ants might be equally common denizens, with antics providing nonstop entertainment. Ants also create exasperation, especially if you've brought a few muesli bars in your luggage. Despite the most foolproof ziplock bag, ants in the Amazon will find and cart off the crumbs to their communities. They find everything edible, anywhere and anytime. As some of our most sophisticated neighbors on the planet, ants create enormous condominiums and assign diverse jobs to different members of the colony. Division of labor and cooperation are requisite traits of ant communities, and they are estimated to comprise 25 percent of the rain forest biomass, meaning there are gazillions of ants. They live everywhere—underground, on bark, in emergent branches, as swarms of aggressive hunters along the forest floor—and sometimes gang up in ant balls to float downriver. Absent from Australia but abundant in the Amazon, leafcutter ants (*Atta* spp.) create highways up the tree trunks to bite coin-shaped leaf bits and haul them down to their underground chambers. They are sophisticated farmers and do not eat the foliage, as one might assume, but create subterranean gardens of special fungus. The fungi break down the fresh leaves, and the ants consume the resulting substance. Go figure! It took entomologists a long time to discover their sophisticated underground farms, which consume 25 percent of a host tree's crown in the process. Citizen scientists often find themselves dancing on the trail to avoid the long lines of leafcutter ants trekking from upper branches to subterranean gardens, hauling green bits up to ten times larger than their own body size and weight.

In addition to leafcutters, other ant species have evolved to live in symbiosis with certain plants. Cecropia (*Cecropia* spp.) is a short-lived pioneer that grows along the river's edge after floods have knocked down existing vegetation. Some cecropia species are an important residence for sloths, which eat its foliage, and others provide shelter for ants living in hollow stem internodes. This symbiotic relationship provides the ants with shelter and food (feeding on the glycogen-containing bodies at the base of petioles), whereas the tree receives protection because ants fight off herbivores and nip away vines encircling the host. I always offer a prize to the first citizen scientist who spots a sloth in a cecropia as we boat downriver toward base camp. Another well-known group of ant plants are shrubs of the family Melastomaceae with swollen chambers on each petiole, which provide shelter to ants serving as bodyguards for the plant. It is disastrous to brush against an ant plant because the residents immediately rush out and bite fiercely. This complex fabric of interactions in a tropical rain forest cannot ever be completely restored after an old-growth or primary rain forest is clear-cut.

Although ants are probably the most abundant creatures in the Amazon, beetles (Coleoptera) are undoubtedly the greatest order of insects, numbering in the millions of species. Some scientists estimate that less than 5 percent of the world's biodiversity has been classified, and a large portion of the remaining 95 percent is beetles. Metallic green, blue polka-dot, phosphorescent pink, bloodred, camouflaged brown or green—beetle coloration is akin to a big box of crayons on steroids! In my lifetime of dangling on a rope, finding beetles feeding on specific foliage has become an obsessive detective game. As I learned in Australia, herbivores often feed at night, so walkways offer a stable platform for spotting nocturnal beetles. After dinner, our citizen science teams don their headlamps and climb into the upper reaches of the forest giants. After twenty-five years, only once did someone step on a venomous snake in the dark. It was a fer-de-lance, coiled in the middle of the trail. About ten people unwittingly walked right over it, until my research assistant, DC Randle, a teacher from rural Minnesota, accidentally flattened it with his large boot. After DC jumped what seemed about twenty feet into the air, our guide spotted the unhappy reptile with his flashlight as it slithered away in shock.

Another nocturnal misadventure involved my son James. When he was eleven, we spent almost two weeks searching for a specific chrysomelid beetle eating a very tough bromeliad along the walkway.

I knew it was a beetle based on the characteristic leaf damage and further deduced it was host specific because almost every bromeliad of one particular species had the same zigzag chewing marks. I wanted to know what in the world ate such amazingly tough foliage. Using their headlamps, James and my Amazon colleague of snake-squishing fame, DC, were monitoring a lush batch of bromeliads near the seventh walkway platform. Suddenly, I heard James exclaim, "Wow!" and then I heard, "Oh no!" He had spotted the beetle feeding, reached out to grab it, and accidentally knocked it off the plant, dropping it some ninety-five feet below into the understory. We were all sad about the incident, but happy to know our quest was worthy—yes, a bromeliad-feeding beetle existed. Within a week, James and DC found another! The host epiphyte (*Aechmea nallyi*), a bromeliad common in the region, was extremely rare throughout the rest of the Amazon. Both beetle and plant are ostensibly endangered species because they are only known to live in one pocket of forest throughout the world, although adequate population data are hard to come by in tropical canopies. We calculated a whopping 10.4 percent insect damage on each bract, not as much as our tree crowns averaging 25 percent herbivory, but an enormous amount for bromeliads with their extremely tough leaves. What an extraordinary feat for the mouthparts of one relatively small beetle! A bromeliad expert once wrote that epiphytes, due to their toughness, were never eaten by herbivores. Never say never! This guy had not climbed into the treetops, where most of the action occurs in a world of plants-defend-against-insects-but-insects-eat-plants-anyway. The tendency for biologists to base their findings at ground level, and overlook the eighth continent, continues to give rise to inaccuracies.

In the world of herbivory, two types of feeding predominate: generalists that eat many different species of plants, and specialists that eat only one species. There are advantages and disadvantages to both. As a human, if you only ate one thing, it could be challenging to move away from the only refrigerator that stocks your food. For host-specific insects in a tropical crown where hundreds of different species coexist, it can be life-threatening if you are blown away from your branch, but if you remain close to your host, it is a comfortable existence. If you are a generalist and eat several types of foliage, then you may have a better chance of survival in a salad bar with so many types of greenery. It quickly gets complicated because there are thousands of shades of green in a tropical forest, each with its own different texture and digestibility, especially critical if you are an herbivorous beetle!

In a dense tropical forest, the canopy sometimes spans upward for several hundred feet. Things change dramatically from bottom to top, but a majority of species reside about two-thirds of the way up. Kind of like the tale of the three bears, this location is exactly right—not too hot or cold, not too dark or light, moderate wind, and ideal humidity. This height receives sun flecks that are ideal for leafing and flowering activities and attracts lots of biodiversity, including the orchid family, Orchidaceae. As the world's largest group of flowering plants, orchids number over twenty thousand species, and a majority are epiphytic. Not all orchid species flower every year, and some only flower once a decade. During nonflowering phases, they escape in plain sight within a sea of green, a great lesson in camouflage. But clever orchid bees— each specific to a single orchid species—manage to navigate to their host plants amid all the greenery. The Greek word *órchis* means "testicle," and orchids are historically associated with love. Tales abound of humans sacrificing their lives for unusual orchids, and exorbitant prices are paid for them. Unfortunately, underworld trafficking of orchids still occurs all too often, and I wish more attention were devoted to saving orchid habitats instead of imprisoning the plants in glass houses. Most of the orchids along our Amazon walkway have been pilfered by poachers who sell illegally to orchidophiles, threatening the conservation of this important group of plants.

In contrast, other organisms in the treetops are less beloved by most of my citizen scientists. Bats are more often feared than loved and zigzag like low-flying but silent Learjets through the forest at dusk and dark. Their consumption of mosquitos is greatly appreciated by field biologists. At our base camp, bats dive-bomb into the outhouses, no doubt munching insects attracted to the odors. It is not uncommon to feel the breeze of a swooping bat on your buttocks, and we joke that they eliminate mosquito bites in certain sensitive places.

Bats are also important forest pollinators, navigating through the mid-canopy via echolocation to pollinate flowers located midway along the tree trunks. Called cauliflory because it looks like cauliflower, such flowers are specific to some species including chocolate (*Theobroma cacao*), which relies on nocturnal bat or moth pollination. Cacao is native to Central and South America, but the largest production comes from Ivory Coast, Ghana, and Indonesia. Worldwide, 90 percent of cacao comes from small family farms of two to five hectares, making it a particularly important crop for indigenous livelihoods.

In addition to bats, many wildlife populations thrive around our base camp, probably because its reserve status has made it a refugium from poachers. The local guides have recently witnessed an increase of mammal sightings, presumably due to reduced hunting pressures in this protected region. Monkeys are abundant! Over time, I have learned to recognize different monkey calls—the high-pitched squeak of the yellow-handed titi, low booming roars of the red howler, and many midrange decibels including woolly, squirrel, and night monkeys, red wakari, monk saki, brown and white-fronted capuchins, saddleback and black-mantle tamarins, and pygmy marmosets. Some of the higher-pitched monkeys and birds go unnoticed by urban visitors whose sense of hearing has been dulled by loud city noises. Despite conservation efforts in the upper Amazon, monkeys are still hunted for the bushmeat trade, especially by logging teams. Mammal poaching leads to an insidious degradation of tropical forests because it is difficult to measure animal population decline as compared to logging. Drones and aerial surveys can easily detect clear-cuts, but they cannot visualize animal deaths. Scientists estimate it will require many hundreds of years for some tropical mammals to recover after overhunting. Sometimes, citizen scientists come across baby monkeys left orphaned after their mothers were killed by loggers or smuggled by poachers involved in the illegal wildlife trade. A conservation effort near Iquitos rears orphaned monkeys, adding to the genetic diversity of local populations when they are returned to the wild. Sloths suffer a similar fate when trees are cut and the mothers killed for bushmeat, leaving the babies vulnerable. Thanks in part to our walkway, conservation is gaining a stronger foothold in this region. Many citizen scientists support conservation efforts when they return home, but perhaps the most important reason to bring citizen scientists to the Amazon is because ecotourism has become an economic driver for indigenous people, providing sustainable income from visitors instead of logging. At the ACTS walkway alone, over a hundred families make a living as boat drivers, cooks, cleaners, guides, shamans, craftspeople, roof weavers, and trail builders. I often tell volunteers that even if you spend the entire week lounging in a hammock, you are nonetheless helping to save the rain forest by providing sustainable employment to the locals so they are not tempted to sell trees as timber. If you also engage in canopy research, then you effectively double your conservation contribution.

In twenty-five years of research at this special place, I've almost never experienced the treetops alone. Volunteers are always keen to accompany me, no matter how hot or dark or rainy. They don't want to miss a minute of action. I only have one memory of a solo walkway visit, unwittingly risking life and limb. I snuck out at midday when it was hotter than the hinges, mostly for a quiet sojourn after a nonstop morning barrage of questions. Sometimes being alone in nature is an elixir. Climbing across six bridges and five platforms surrounded by tropical humidity from so much transpiring foliage, I wiped the sweat dripping off my nose and lost the battle to prevent my glasses from steaming up. In the middle of a sultry day, even mosquitos had taken refuge. I'd been sitting on platform number six for two straight weeks of fifty-three broadcasts during the JASON Project, so this vista had become a second home, where problems fall away. I knew exactly when the treefrogs would start singing, which bromeliads housed a tarantula, and what branches on the horizon hosted morning meetings for toucans. Standing at the top of the world, in the upper limbs of the green bean tree, my problems did indeed fall away. This was always a spiritual sensation, and sucking in the fresh air all by myself was an incredibly special tonic. In the distance, a screaming piha shrieked, possibly announcing its discovery of a snake or lizard coiled in hiding. I watched in awe as a flock of scarlet macaws flew over, probably on a mission to find fig fruits. Then I spotted an ominous black cloud looming up in the west, so maybe the birds were seeking shelter? In a millisecond, lightning struck a nearby tree with a hellish crack. By some miracle, it did not hit my giant emergent. We have a rule in the canopy—immediately go down to the ground if thunder and lightning strike. But I felt transfixed and stood still in this aerial roost. Within two minutes, the sky went from sultry sun to dark thunderclouds and swirling wind gusts. Rain lashed around me, drops pelting fast and furious. After a few seconds of fear, I was thrilled to experience my favorite place during a powerful storm. With the treetops illuminated by lightning, I felt exhilarated to be soaking wet like all the leaves around me, watching branches dance wildly in the wind. After the brute strength of the storm, I felt cleansed and ready to return to the constant onslaught of questions posed by eager citizen scientists. Once in a while, it feels good to stand still and absorb the ferocity of Mother Nature.

Except during the wildest thunderstorms, walkways offer safe access for arbornauts. During expeditions, volunteers visit the canopy at all hours of the day, during different seasons and weather, and for extended periods of time, measuring herbivory on dozens of trees, vines, epiphytes, and hemi-epiphytes. In my years of research there, herbivory ranged from as low as 1.8 percent leaf surface area eaten in philodendron, to 10.4 percent for the bromeliad *Aechmea*, to approximately 30 percent for *Inga* sp. But scientists hypothesize climate change will jeopardize the future of forest canopies, causing hotter temperatures and higher frequency of drought, both of which are ideal for insect outbreaks. So insect damage is predicted to increase in tropical forests. What is the meaning of "high levels of herbivory" other than a human judgment about holes in leaves? How do we know if insect frass raining from the sky by herbivores might outweigh the negative aspects of losing a bit of green tissue from this aerial salad bar? Nutrient cycling from sunlight to leaves to insect poop to soil to roots and back into crown growth is still a mystery. The fate of forest canopies with the onset of climate change remains unknown, despite their critical importance to planetary health. The high humidity and lack of electricity, plus the absence of a controlled laboratory, dissuade most scientists from conducting field research in places like Peru, Ethiopia, or Cameroon. As a result, global ecological knowledge is not evenly distributed among geographic regions, with a preponderance of tropical forest research undertaken in Mexico, Panama, Costa Rica, and Brazil, in part due to the attraction of well-funded, well-appointed research labs. Until we achieve more equality in the distribution of scientific tools and intellectual capital across ecosystems whereby biologists work in all important habitats and not just convenient ones, we may never truly unravel the complexities of many underexplored treetops.

For citizen scientists, wildlife encounters in a remote tropical jungle are reminders that we are a long way from Walmart and fast food. The word "Amazon" takes on a new meaning when our lives depend on a river, not an internet shopping site. Volunteers are asked to try a hand at living off the land by fishing for piranha using local sticks as poles. During one boat trip, we caught thirty-one fish—but all so small they only made tiny appetizers. Another time, the team caught zero fish and would have gone hungry if not for the kindness of local villagers sharing their catch. Once, the gawkiest and most insecure college student caught a record-sized piranha! It was a wonderful moment for her. I thought back to my fifth-grade science fair, when I was awarded that plastic second-prize trophy for a wildflower collection, and knew this fish was her parallel moment of glory.

For most of the year, tropical trees look alike with elliptical, entire leaves and smooth brown bark. But the local shaman knows every species and has learned about important plants over many generations. In contrast, professional botanists often train for a lifetime, painstakingly waiting years for tropical plants to flower and fruit, because sometimes those are the only features that differentiate one species from another. Identification can't be mastered during one quick expedition. When I take groups of people into the canopy, I ask them to think about the analogy of visiting an art gallery. If you see one or two painters you recognize, suddenly you feel "at home." It is the same in the rain forest. If you find the pixie cups of *Inga* or the sea-urchin fruits of *Apeiba*, then suddenly you have friends among an otherwise homogeneous-seeming salad bar. But it is impossible to learn too many trees at once, because they look so much alike to the amateur eye—green, oval, toothless leaves.

One practical use of tropical vegetation by ribereños in the upper Amazon is the construction of blowguns, silent weapons that allow sustainable hunting by these indigenous river-dwelling families. First, the shaft is carved from the blowgun tree, called pucuna caspi by the Yagua Indians. They cut a long section of wood in half with a machete, and then whittle a channel out of the center of each piece. When two opposite sections are aligned and carved, a tar-like plant resin is used to cement them back together, leaving a hollow tunnel through the center. The flattened aerial roots of a philodendron are wrapped around both halves to seal the blowgun, and a mouthpiece is carved from the lightweight timber of local mulberry. The darts, carved from palms, require a toxic substance called curare to tip the arrows for killing game. Made from the bark of a liana (*Curarea toxicofera*) and a shrubby tree called strychnos (*Strychnos panurensis*), curare reduces neuromuscular activity in the prey, resulting in paralysis and eventual death from respiratory failure. Great kapok seeds are propelled by a silk cotton material, and this substance is harvested to wrap around the blowgun dart, creating the aerodynamics to propel it through a six-foot-long blowgun chamber. The darts are silent, quick, and deadly to a monkey or capybara, and over many generations the local people have hunted sustainably without endangering local wildlife populations.

When Eddie and James were only eleven and ten, respectively, they accompanied me to the Amazon during a Christmas holiday. There was never any plan B when their mom traveled since, in a single-parent household, the boys could not stay home alone and I tried not to over-utilize the favor bank of their grandparents. James felt sorry for himself because he missed his first-ever invitation to a New Year's Eve party. But then the village shaman pulled him aside and offered to teach him how to use a blowgun. Not only was he a deadeye shot, but he got his very own instrument made by the Yagua Indians. After returning home, he never thought once about missing that party, and proudly took his blowgun to school (without darts) and explained its botanical parts to his fifth-grade science class.

I love bringing entire families to the rain forest, because children often become the best citizen scientists, honing their five senses more quickly than their parents. In Belize, my children found a new species of slingshot spider busily hunting food by zinging a silk strand plus itself forward at lightning speed to snare prey. Arachnologists later measured the acceleration of a slingshot spider's silk strand at 1,200 yards per second squared (as compared to a cheetah, which chases prey at a mere 13 yards per second squared). Both James and Eddie had eagle eyes not only for finding spiders hanging under branches or tiny beetles feeding on leaves, but also for locating well-camouflaged pink-toed tarantulas hunting prey on the edge of bromeliad tanks. As children, they had a pet tarantula, easier to manage than dogs or cats given our frequent travel schedule. At first, it was named Harry due to its distinct hirsute exoskeleton, but when it grew enormous, the boys realized it was female, so the spider was renamed Harriet. In the spider world, females can be up to a hundred times larger than their male counterparts. Every two weeks, we dropped one cricket into Harriet's glass terrarium, and she stalked, terrorized, and finally pounced on her prey. The neighborhood kids loved to watch her feeding exploits. If you blinked, you might miss the rapid hunting prowess. What an ideal pet, which only required a single cricket every two weeks. Pink-toed tarantulas were relatively common in the Amazon, living up near bromeliads where they captured insects coming to sip water or lay eggs. But they also set up housekeeping in the rafters over our beds, hunting insects that are unfailingly attracted to humans. One year, when one citizen science family arrived with four teenage daughters, there was an enormous shriek when they spotted a tarantula in their bunkroom. Knowing this was a make-or-break moment for the success of the trip, I instantly praised the teenagers for having the coolest room and making the first observation of this awesome species. They kind of looked at one another wide-eyed, and soon after bragged to all the other kids about their amazing eight-legged roommate. They went on to have a life-changing expedition, with family memories and science exposure no school classroom could ever offer. This same family held a boa constrictor several days later. That must have provided an unforgettable family holiday photo!

Despite long-term research by many dedicated scientists, the Amazon is in big trouble. As the world's largest remaining rain forest, its future is uncertain—clearing, burning, logging, and roadbuilding are on the rise. The estimated 6.2 million square miles (or 4 billion acres) of original primary rain forest has been halved to less than 3.5 million square miles in 2020. Most of that deforestation happened in my lifetime, primarily due to the North American appetite for beef, soy, palm oil, and other agricultural products, and the onset of road construction has led to a frenzy of gold mining and oil drilling, which destroy the forests and create toxic pollution. During one year from mid-2017 to mid-2018, deforestation of the Amazon rose 13.7 percent. Even worse, throughout 2018 the Brazilian Amazon saw an estimated 200 percent increase in deforestation from the prior year, according to the biologist Antonio Donato Nobre (*Climate News Network*, March 16, 2020). And since 2018, rain forest conservation regulations have been very lax under the Brazilian president Jair Bolsonaro, including massive burning. According to Nobre, land grabbers organized a "day of fires" during August 2019 to honor Bolsonaro and his disregard for the value of the Amazon rain forest. Such significant losses of the eastern Amazon in Brazil have increased the value of the western, and less degraded, rain forests in Peru. Tropical rain forests cover less than 10 percent of Earth's landmass but house approximately two-thirds of the world's terrestrial biodiversity, with a significant majority of those species in the canopy. The Amazon rain forest took fifty-eight million years to evolve, but scientists predict it could pass a tipping point within the next fifty years, collapsing into dry savanna because an excessive loss of foliage will preclude the normal rainfall patterns. Nobre and his colleague Tom Lovejoy of George Mason University predict that if Amazon deforestation exceeds 20 percent, the hydrological cycle will cease to provide enough rainfall to support forests (as well as humans). In addition to their importance for global moisture circulation, rain forests absorb carbon dioxide (which humans have emitted as pollution), and approximately half the dry weight of every enormous, old-growth trunk represents carbon storage. When forests are burned, the fires release carbon back into the atmosphere. And when the Amazon is cleared, rainfall decreases significantly in the absence of leafy canopy that served as a recycling agent for moisture.

Conservation of rain forests requires an informed and participatory public. Citizen science has become unquestionably successful in fostering public outreach, not only in museums but also for NGOs, state governments, local policy makers, and K–12 science. This is a game changer in the current political climate, where much of the public increasingly casts doubt on science due to misinformation. Expanding environmental literacy is vital in the twenty-first century so our growing global population understands the limits of natural resources. We are fast approaching tipping points where many ecosystems, including the Amazon, reach irreversible damage. Yet, never before have humans had such a wealth of technology to innovate solutions, including the capacity to collaborate from virtually anywhere around the world, draw ideas from multiple disciplines, analyze countless data points, and create novel toolkits for STEM. Planetary scientists must seek to balance cellular versus organismal biology, virtual models versus real-time data, and science blended with policy. Future stewards need skills to assess, predict, manage, and communicate the ecological and societal changes emerging from a dramatically altered global landscape. However, a major stumbling block in training the next generation of practitioners is how to effectively integrate virtual technology with in situ fieldwork. While most older ecologists were originally inspired by outdoor play, younger scientists interact with ecosystems through virtual gaming, social networking, and computer models, sometimes leading to "nature deficit disorder" whereby kids stay indoors. The author Richard Louv, in his bestselling book *Last Child in the Woods*, reported one youth exclaiming, "I wanna play indoors, 'cause that is where all the electric outlets are." So, how can environmental practitioners blend hands-on fieldwork with virtual technology? This conundrum is the subject of ongoing debate, but citizen science is one of the creative solutions.

For any country to retain global competitiveness, it needs to encourage STEM innovation and education. Research investments in China, Singapore, and South Korea are topping the charts for scientific literacy of their students and citizen scientists. America increasingly lags behind and, according to the National Academy of Sciences (NAS), spends more on potato chips than the federal government's budget for research and development of energy. The NAS also reports that only 4 percent of Americans work in science and engineering, but this group creates jobs for the other 96 percent. When scientists develop a new diagnostic tool for cancer or engineers patent clean energy technology, these innovations translate into jobs in manufacturing, marketing, transportation, sales, and maintenance, as well as education and training. In recent history, STEM innovations have dramatically transformed the way we live. For example, tape recorders were replaced with iPods, maps with GPS, landlines with cell phones, two-dimensional X-rays into three-dimensional CT scans, and slide rules and daily planners became computers. But an enormous roadblock still exists: some fourteen thousand American public school systems are suffering declines in student proficiency in math and science.

Citizen science is part of a broader solution to reverse these trailing STEM metrics. Getting kids involved in bird counts, shoreline trash pickup, local BioBlitzes to conduct rapid species counts, urban tree planting, or water quality testing is a good start. Thousands of citizens are using an app on their mobile phones called iNaturalist, which compiles photos of biodiversity and maps their distribution. Galaxy Zoo is another computer-based imagery system where citizens search for stars, galaxies, and other extraterrestrial sightings on authentic NASA photos. Other emerging programs illustrate the integration of citizen science with technical scientific research. The National Ecological Observatory Network (NEON) is a twenty-first-century National Science Foundation project that conducts continental-scale environmental monitoring with large databases accessible to students, citizen scientists, and policy makers. I was one of sixteen scientists who wrote the $300 million-plus grant to NSF to fund the NEON platform. After several years of debate and strategic thinking, our committee reached consensus that such a major initiative would generate long-term data to better understand global change and ecosystem responses as well as engage diverse audiences through many platforms. Similar initiatives are underway in other countries, where citizen-assisted monitoring can gather valuable data. Singapore has invested over $20 million in a vast series of walkways through urban forests, providing incredible access for birding, phenology, and insect counts. In the museum world, citizen science and public outreach are watchwords of success. When I was the director of the Nature Research Center in Raleigh, North Carolina, we partnered with North Carolina State University to swab belly buttons, cultivating the bacteria in petri dishes, and giving citizens some insights into their own body's biodiversity. Some scientific questions can be answered more comprehensively from a multiplier effect when the public is involved.

But new tools and technologies alone will not conserve ecosystems or save species. Educated citizen scientists as part of a broader public can do almost anything from finding insects in the Amazon rain forest to mapping human belly button bacteria to counting leaves. Especially after the COVID-19 pandemic, many teachers now have expanded to a classroom education process unbounded by walls, where handheld technologies such as iPhone applications are increasingly available for science education. The big challenge is not a lack of information, but articulating a relevant context such as insect outbreaks or urban canopy cover that will motivate future generations to embrace ecological stewardship. Linking healthy ecosystems to economics and human health is one important stepping-stone. But amid all the technology, students still need curiosity and a thirst for discovery. This does not require expensive equipment, just a chance to engage their five senses by playing outdoors. If students and citizen scientists develop curiosity about nature, not just pushing buttons on video games, then it seems certain they will be more likely to solve the grand scientific challenges of the near future.

In 2022, I hope to lead my twenty-fifth citizen science expedition to the Peruvian Amazon, where volunteers will continue to explore tropical forests using the world's longest walkway. Despite nearly complete immersion in nature, over the years we've sometimes witnessed rafts of logs clogging the waterways, even as far upriver as Iquitos. An oil refinery is now anchored on a barge just several miles from our skywalk base camp, having sailed over two thousand miles from the mouth of the Amazon with all its hardware and toxic chemicals. And next door in Brazil, extensive fires burned over 6,700 square kilometers (4,160 square miles) of rain forest between January and August 2020, releasing 225.8 million metric tons (MMT) of emissions, according to a paper in *Science* magazine. When the Amazon burns, it not only emits carbon and destroys critical habitat for millions of species, but it also results in air pollution that exacerbates human health. And such deforestation only reflects that portion clearly visible by aerial mapping, not insidious degradation such as roads, selective logging, or edge effects that are tougher if not impossible to measure without ground-truthing (i.e., up-close human reconnaissance). Wistfully, I've reminded my citizen scientists that they are privileged to experience this most beautiful tropical roof of the world, because if we don't do something dramatic as a species to turn the tide on climate change, it will very soon be gone.

⫸ The Great Kapok Tree ⫷
(*Ceiba pentandra*)

DURING FREQUENT TRIPS ON AMAZON RIVERBOATS, one lone kapok (*Ceiba pentandra*) towered over the riverbank about five miles outside of Iquitos, Peru. Each July, when bringing citizen scientists to the walkway, I gazed in awe from the boat at this handsome silhouette. I wanted to ascend that tree, and eventually persuaded a climbing friend to accompany me. We were informed by locals the whereabouts of a shaman who had jurisdiction; it was not respectful to climb without her permission. So we hesitantly knocked at her hut and asked. She was more than friendly and said, without any specifics, if the spirits were willing, we could climb. Hoisting our longest ropes over our shoulders, we trekked to the base of this giant, wondering what she meant. It was much bigger up close than it appeared from afar. Like most kapoks, a single trunk rose at least a hundred feet, and then another hundred feet of hefty horizontal branches angled straight out from the main trunk, almost like aerial shelves. There was only one possible shot over the first branch in between festoons of vines and epiphytes. We used our biggest slingshot, called the Big Boy. It stands on the ground and consists of a vertical three-foot pole that serves as a launchpad for propelling the line. We both gritted our teeth, grimaced, and pulled back its giant elastic band. *Crack!* The rubber snapped loudly, and our fishline soared up and out of sight. We squinted, and I used my binoculars. A perfect shot! The Amazon spirits wanted us to climb. I was first, but close to the top and seemingly almost halfway to heaven, I felt a quiver on the rope. What was going on? Looking aloft, I saw an enormous bird on the branch, pecking the rope. It was a horned screamer, and we later learned they roost in these emergents. Obviously, the rope looked like a snake, so the bird was defending its aerial perch. I hustled up to prevent my lifeline from being severed and gently shooed away the handsome screamer, which reluctantly shared space with me. Once aloft, I saw that the kapok crown was a lush garden of bromeliads, orchids, philodendrons, buzzing insects, and vines snaking every which way. I collected leaf samples of the most common plants to calculate their herbivory but never climbed that tree again, as a personal homage to its seemingly unattainable stature.

Mayan mythology in Central America claims *Ceiba* is an important "tree of life" and represents a universal communication between the three levels of earth (underworld, middle world, and upper world). Its roots represent the underworld, the trunk is the middle world inhabited by humans, and the canopy symbolizes the upper world. Kapok trunks bear thorns during juvenile stages, and they are often portrayed on Mayan pottery as symbols of reverence. Adults develop buttresses instead of thorns, and such towering emergents are visible for tens of miles on the horizon. Unfortunately, their stature also pinpoints their location for loggers. Most kapoks have been cut and the Amazon skyscape reveals almost no remaining individuals, except one or two protected by a local shaman for spiritual and medicinal purposes.

The kapok is a veritable apothecary in the sky. Its leaves are used against scabies, diarrhea, fatigue, and lumbago, as a laxative, and to combat heart ailments. The sap is drunk to treat mental illness, stillness of limbs, fatigue, headache, coughs, and eye wounds, and is administered to children to prevent stomach parasites. The trunk is boiled to treat toothache, stomach ailments, hernia, gonorrhea, edema, fever, asthma, and rickets, as well as injuries such as wounds, sores, and even leprous macules. Bark extracts create an effective enema, and tender shoots make a contraceptive. Finally, the roots are equally medicinal, curing diarrhea, dysentery, leprosy, and hypertension. I guess it was no surprise we had to ask permission of the shaman to climb a kapok, since this tree essentially serves as her drugstore in the forest.

Ceiba is derived from a Carib word for a dugout boat; and *pentandra* is Latin for "five-stemmed," which refers to its compound leaves composed of five leaflets. Although allegedly evolved in the American tropics, records of kapoks in Africa call this into question. The genus consists of fifteen species, each palmately compound and deciduous, losing their foliage for several months during the dry season. They flower before leafing, and their creamy white to reddish flowers open after dark, emitting a strong, unpleasant odor that attracts bats to drink the nectar. The bats carry pollen on their fur and transfer it to other flowers, facilitating pollination. Moths are additional night pollinators, although timing and location are critical because not all individual trees bloom each year.

Kapoks belong to the Malvaceae family, which houses approximately 25 genera, formerly called the Bombacaceae family. Individuals grow to 250 feet high, the tallest spires of the Amazon. In other countries, different *Ceiba* species serve as meeting spots, shade for coffee crops, sacred trees near temples (in India), and shelter for livestock. A white cottony fluff (which provides aerodynamics for blowgun darts) disseminates the kapok seeds and is the source of the common name, silk-cotton tree. *Ceiba pentandra*, the predominant species in upper Amazonian Peru, is also called lupuna, ceibo, or ceibote. Kapok cotton is still harvested in Asia and Indonesia, although artificial materials substitute for the stuffing formerly used in life jackets and mattresses. Due to their elegant form, many species were cultivated and hybridized by indigenous people throughout South America before the arrival of Europeans, making identification difficult.

In my small apartment, I have a special wall of blowguns, each of which represents the memory of an Amazon adventure over the past twenty-five years of field expeditions. Sometimes I arrive home from another long expedition to find the front hall snowing with kapok silk that escaped from the blowgun pouches during my absence, a wonderful reminder of this extraordinary and useful tree that represents the overstory of the Amazon canopy.

8

TIGER TRACKS, TREE LEOPARDS, AND
VEDIPPALA FRUITS
·»)·((·
Exporting My Toolkit to Train Arbornauts in India

W E HAD DRIVEN OVER EIGHT HOURS south from Bangalore, India into the remote Western Ghats to work at a
field station dedicated to forest ecology in India's highest
biodiversity forests. Covered with dust from the drive and exhausted
from the chaotic roadways, we nonetheless donned our hiking boots
to get a quick look at the dominant vedippala (*Cullenia exarillata*)
tree canopies of this region. I had heard about these trees, with their
keystone role as providers of fruit for many food chains in the forest,
because two of my colleagues, T Ganesh and Soubadra Devy, were
experts on their natural history. While hiking away from our rustic
camp to get my first glimpses of India's subtropical forest canopy, I saw
T Ganesh suddenly leap in excitement and shout, "Tiger—BIG tiger!"
My heart pounded. No other sounds were audible except the rustling
branches of vedippala at least one hundred feet overhead. He pointed
to an enormous footprint, freshly imprinted on the sandy soil. The
track was significantly larger than a human hand, and my colleague
was breathless because he determined this cat had walked ahead of
us and had probably watched us climb up a tree to observe lion-tailed

macaques only minutes before. Nearby we found claw marks on a trunk, clues that this beast was not only large but also we had trespassed in his territory, inciting him to scratch the bark to mark his dominance. We practically tiptoed home, fearful yet also anxiously hoping to glimpse a "king of the jungle."

The Kalakad Mundanthurai Tiger Reserve (KMTR) in the Western Ghats of southern India was officially one of approximately fifty biodiversity reserves for India's fifteen hundred remaining big cats. This reserve also boasted 3.3 percent floral endemism, meaning that the plant species in this place are found nowhere else in the world. It provided a refuge not only for tigers and plants but also for highly endangered primates, civets, bats, and many species of birds, butterflies, amphibians, and reptiles, all of which depend on the health of the canopy. Yet the nine-hundred-square-kilometer (560 square mile) reserve was threatened by encroaching land-use pressures, including tea and coffee estates, rice paddies, and rising incidents of poaching. A few months after that first encounter at KMTR, eight tigers disappeared from the nearby Ranthambore National Park, victims of poachers. Despite their endangered status, tigers are still hunted for their skins and body parts, some of which constitute ingredients of traditional Chinese medicines thought to enhance sexual prowess. The world is well aware of the endangered status of tiger populations but less aware of the widespread clearing of the actual forests they inhabit. As arbornauts, my two Indian colleagues and I were not directly studying large mammal populations, but we documented the health of their canopies overhead. It is critical to save species, but that requires saving their habitat. This was the reason I was here—to foster conservation through international collaboration, which translates into saving whole forests, from the bottom all the way to the top, and making my arbornaut's toolkit available to share with everyone.

Field biology is shaped by time and funding. Opportunities to conduct long-term, continuous research are the toughest to obtain. In Australia, I had the good fortune to receive three years of a graduate fellowship to collect long-term data on leaves. My Amazon research involved short time frames of data collection because the citizen science expeditions operated for only ten days, based on volunteer availability. As a result, I designed research questions that could be answered by brief visits, though I did have the good fortune to return to the same trees over many years. A third type of fieldwork, which is the toughest to design for useful outcomes, involves an occasional, once-only trip to a site. I call this "snapshot research." To achieve professional goals in these kinds of circumstances, I often turn to outreach activities, including mentoring women in science, conducting rapid surveys such as BioBlitzes, presenting public talks, and training students. Such snapshot activities represent new ways of applying science, a shift from the conventional research of doing one thing in one place for an awfully long time. The rapidly changing planet has galvanized the scientific community to operate quite differently from the past in order to achieve bold conservation aims.

In 1994, I organized and hosted the first international canopy conference at Selby Gardens in Sarasota, Florida, to address the challenges of global collaboration in both forest science and conservation. I searched the recent literature to identify a handful of India's emerging tree experts as potential invitees and found funding for four of them to participate. Two of those colleagues, T Ganesh and Soubadra Devy, were a devoted married couple who had dedicated their lives to the exploration of India's eighth continent. Another, Pallaty Sinu, later collaborated with me on India's sacred groves. At that first conference, 250 arbornauts from twenty-five countries were dumbstruck by the realization that most of us had never met one another. We had literally spent the last few decades dangling from ropes in remote locations, sailing over forests in hot-air balloons, building walkways, or frantically counting millions of new species in those lofty heights. The conference was a huge success, leading to increased global partnerships and sharing of our methods among colleagues. Four years later, I hosted a second international conference in Sarasota, this time attracting arbornauts from thirty-five countries and featuring a public exhibit of new treetop tools including the inflatable raft, specialized tree sleeping hammocks, sleek and improved slingshots, and lightweight climbing hardware. Funny to think that the collaboration of international canopy researchers unofficially premiered in Florida, a state where forests are usually cleared for shopping malls and golf courses. The third and fourth meetings were hosted by canopy scientists in Germany and Australia, respectively.

Several years after meeting Soubadra and T Ganesh at the inaugural conference, I visited India as a Fulbright senior scholar, with a mission to ramp up treetop research and discuss options for a public walkway with India's foresters. I was based at Ashoka Trust for Research in Ecology and the Environment (ATREE) in Bangalore, a city whose "canopy" was a profusion of electric wires, not tree branches. Not surprisingly in a country with such rapidly rising population, as technology expanded, nature contracted; the wires and cement were fast outpacing the green spaces. All roads to and from Bangalore typically bustled with bicycles, trucks, and cars interspersed with oxen, cattle, and donkey carts. The delicious aromas of curry intermixed with fumes of diesel. On my very first day, a statewide strike closed all businesses, including restaurants. Because I was hungry after the time change and long flights, Soubadra miraculously managed to find a friend who owned a small shop and convinced him to serve us lunch. We ate quietly in his back room so no one could see. He did not have much food in stock, because shopping at the market was a daily activity for any Indian restaurateur. No wine, no milk, no ice water, no cheese, no fruit, no meat, no salad, but a delicious mixture of rice, spices, vegetables, and broth! When the strike ended the next day, streets resumed their normal chaos, and stalls reopened selling absolutely everything: live chickens, plastic containers, food, firewood, blue jeans, lumber, wire, palm readings. Over many years and many return visits, I have witnessed India metamorphose, like a caterpillar changing into a butterfly, or perhaps a proverbial seedling transforming into a canopy tree. My diaries include vivid descriptions of lying in hostels all night long listening to dogfights, using water jugs instead of toilet paper for ablutions, undertaking terrifying taxi rides that careened along roads strewn with cattle and carts, and challenging my Western digestive tract with food served in roadside stalls. The biggest take-home lesson from working on forest conservation in India was to learn patience and allow for longer time frames in getting things done. An eight-hour taxi ride to a field station after a long plane flight was commonplace, as was a two-week delay in getting a permit to enter a forest. Fellow arbornauts Soubadra and T Ganesh were always extremely apologetic when we had to wait for things or traffic jams delayed our agenda, but I actually loved seeing all the sights and sounds during such delays as we made our way from urban to remote destinations. One day, we took a taxi that proceeded to drive along the sidewalk, the driver honking at bicycles and oxen in his path. While my hosts cowered in their seats, I was totally mesmerized by the driver's ability to do the seemingly impossible. And we arrived on time.

Even the role of women in science in India has advanced rapidly, although not as fast as it should. Soubadra was the first and still one of the only female canopy biologists in the entire country, a ratio lagging behind the United States and Europe where women comprise closer to 10 percent of canopy scientists. I admire her pioneering spirit so much that I nominated her for the famous Lowell Thomas Award at the Explorers Club a few years ago. It was one thing to study the treetops of Australia or California, but a whole different level of challenge in India where both infrastructure and research support were not reliable. During my tenure in India as a Fulbright scholar, I gave a talk to the forestry department in Kerala, capital city of Thiruvananthapuram. Forty-five foresters, all men, sat formally behind enormous mahogany tables, with a microphone at each place, while their director, Shri. N.V. Trivedi Babu, headed up the group. After a brief blackout during which we all sat calmly awaiting electricity, I projected my presentation on a minuscule screen, at which they all squinted. I had taken Soubadra's wise advice and worn a traditional sari, so I at least looked somewhat acceptable to a roomful of senior male foresters. Despite the challenge of the small slide images, the men warmed to my message about the critical importance of forest canopies and became enthusiastic about building a skywalk in India. T Ganesh told me later they were all quite surprised, even speechless, that a woman was leading their discussion. At that time, India faced a conundrum: forests were considered a government resource managed by professionals who guarded them, but no plan existed to share forest reserves with a greater public. So trees had become an asset that was locked up and off limits to almost everyone. Ironically, over the past two hundred years, America has cut down a whopping 97 percent of her primary forests, whereas India has retained 21 percent—an incredible treasure. So most of India's trees remained behind locked gates, and the government foresters were rightfully concerned that ecotourism would irreversibly degrade them. Despite wanting people to access their country's natural resources, I could sympathize. America's Yellowstone National Park is overrun every summer and the United States has only 330 million people. Giving India's billion people access to its precious forest reserves would undoubtedly prove difficult to manage. The foresters asked excellent questions about the prospect of constructing an aerial trail in India: Do walkways harm wildlife? How many people can a bridge hold? Are there ways to limit access? How about signage? We achieved my Fulbright goal that day—inspiring India's forestry leaders to think seriously about the notion of sharing their trees with their public. Although I would have loved to see a skywalk instantly embraced and constructed, I have learned that conservation often advances with tiny steps, especially when shared across cultures. We are still working toward a canopy walkway in the Western Ghats, a project that may soon be close to fruition.

After the meeting, Soubadra and T Ganesh took me to a local wetland down the road for birdwatching. They shared my lifelong love of feathered creatures. Thrilled, I added five new birds to my life list: redwattled lapwing, oriental pied hornbill, Indian river tern, black ibis, and openbill stork. Birds remain a passion, ever since those childhood bird egg collections and the magic of holding a goldfinch while bird banding at the summer wildlife camp.

We were sure that engaging international partnerships could jump-start forest research for Soubadra and T Ganesh, so we three agreed to dream big and fundraise to cohost the fifth international canopy conference in Bangalore. This was a huge debut for a country with only a few professional arbornauts, but it was also home to primary forests full of endemic, relatively undiscovered species. It would be the first time our meeting was held in an economically emerging country, a healthy shift for a growing network of professionals interested in making a difference across the globe. Part of my mid-career strategy was to prioritize working in countries that were underserved in terms of scientific resources—human capital and funding. Just as we seek to expand STEM training for diverse populations such as women and minority students, the scientific community similarly needs to seek equity in the distribution of research dollars and intellectual capital throughout the world. India is one of the countries where the practice of forest science has lagged behind North and South America. It was a perfect petri dish for seeking conservation solutions, but could benefit from a dedicated interface of local and international scientists.

We three wrote several conference grants to host distinguished speakers and also relied on the good nature of colleagues for donations of their time, equipment, and talent. Our budget did not stretch enough to provide everyone with travel reimbursements or per diem expenses, but we hoped our fellow arbornauts from industrialized countries would graciously participate. For field biologists, international travel and conferences literally bring folks "down from the trees," sharing cultural differences that ultimately lead to greater respect and trust. Conferences are the conventional mechanisms for scientific information sharing, but some folks need cajoling with airfare or the promise of a published volume to disseminate their findings. We tried all of the above—funding travel for the most distinguished speakers (which has a trickle-down effect of inspiring others to attend), finding a publisher for a proceedings volume, and then creating a lively agenda to entice students and young professionals. We encountered a few humorous issues, such as the hotel having excessive mosquitos and some staffing challenges, but we more than compensated with activities such as tree-climbing workshops and delicious Indian banquets. Yes, there were crazy stories about flights, arrivals, taxis, and even the hotel—but it contributed to the legendary experience of the meeting.

The fifth international conference in India, entitled "Forest Canopies: Conservation, Climate Change and Sustainable Use," served specifically as a wake-up call for forest conservation, with five specific spotlights:

1. Deforestation—As canopy science is unfolding, we have gained the ability to map clear-cuts with greater accuracy. New tools such as LIDAR (Light Detection and Ranging, which is a new imagery technique that provides detailed forest signatures such as information about tree carbon or water content), satellite technology, and other aerial reconnaissance enabled us in 2009 to calculate the sobering disappearance of some thirty million acres of rain forest that year. Even more tragic: approximately 80 percent of timber removal from the Amazon was illegal, and usually made its way into North America through harbors that do not reliably monitor imports.

2. Amazon basin—Scientists modeling climate change predict the tipping point for irreversible damage to Amazon tropical forests, and subsequent catastrophic impact on our planet's weather patterns, is estimated at 20 percent degradation. Our keynote speaker, Tom Lovejoy, pointed to aerial surveys indicating it had already reached a level of about 17 percent by 2009. (Footnote: alas, it subsequently exceeded 20 percent within the next decade.) Forest canopies—and the relatively new research approach of studying whole forests, not just the base of a tree— were acknowledged in this meeting as a critical conservation priority for the future of planetary health.

3. Carbon storage—Recent whole-forest research has led to the discovery that trees represent an important sink for carbon storage, since big trees in particular absorb a large amount of carbon dioxide from the atmosphere. We talked about old-growth or primary stands as a new type of global currency. Scientists and economists have recently proposed industrialized countries with large carbon dioxide emissions should pay tropical countries to keep their forests intact as an important mechanism to

reduce climate change, because foliage removes an estimated 20 percent of carbon dioxide from the atmosphere.

4. Biodiversity—Canopy science inspired an early twenty-first-century "green rush" to discover species, since a majority lived in the tops of trees, and India was a prime example of a relatively unexplored territory. Until arbornauts began accessing the treetops with the advent of ropes and walkways in the 1980s, this biodiversity remained uncharted. Some of India's charismatic creatures were highlighted at the conference, including the Indian cobra, flying lizard, Malabar parakeet, great hornbill, lion-tailed macaque, Nilgiri langur, and Indian tiger. At that time, it was estimated that over 90 percent of India's biodiversity remained unclassified in scientific records. Our conference jump-started attention on India's unexplored forests, and the media featured our pronouncements with big headlines.

5. International collaboration—The future of the world's trees will require teams of field biologists working together, not solo. The India conference was a turning point because such a gathering of experts inspired future partnerships. All the arbornauts resolved to return to their respective countries to jump-start field research, to help link economics and science through an understanding of the services that forests provide to ensure human health, and to share their newfound discoveries about the treetops with a diverse public and with appropriate leadership.

The first four international conferences inspired discovery, inspiration, and methods to study forest canopies. In addition to those priorities, the fifth meeting in India communicated for the first time a clear state of emergency about the state of global forests. Canopy access tools had now been in operation for almost thirty years and spread to over thirty-five countries, so the integral links between forest health and planetary sustainability were increasingly obvious from the results of ongoing fieldwork. In short, forests were worth more alive than dead; carbon storage, water purification, soil conservation, productivity, and homes for biodiversity were now allocated monetary values, not just timber harvest alone. Trees were fast becoming a new planetary currency. Around the time of our conference, the Rocky Mountain Institute in Colorado calculated a "natural capital" value of forests at $4.7 trillion, and even this enormous number probably underestimated their services.

The conference in India spotlighted a sense of urgency for all arbornauts to work tirelessly back in their home countries. Within India, the national media shared our scientific findings, garnering headlines throughout the country. The *Deccan Herald* published a feature article, as did the *Bangalore Mirror, The Hindu,* and *The Times of India,* which included a photo legend cleverly and heartbreakingly labeled "Withering Heights" to emphasize the demise of the country's forests. Thanks to India's journalists and the popularity of newspapers, over twenty-five million people learned about the importance of canopies. I smiled to see a dozen Indian men squatting around one newspaper on a local sidewalk, discussing its current events.

That meeting debuted two new advances in the canopy toolkit: (a) aerial surveillance using LIDAR, drones, and satellites, and (b) an increased number of women joining the ranks of field biology and conservation decision-making. Drones and other aerial imaging saved time, energy, and perspiration. Counting the number of flowering vines or mapping primate nests with drone imagery avoids the need to climb dozens of tall crowns over many weeks. As technologies such as LIDAR become more precise, scientists can analyze drought, insect outbreaks, and many other attributes of forest vitality because the advanced imagery technically analyzes foliage moisture, stand health, or even carbon storage. This type of information was invaluable to forest management, especially when combined with ground-truthing. Why do we still undertake ground-truthing, which requires someone to climb and look at crowns from up close instead of from afar, when aerial surveillance can reduce the physical effort and perspiration? It is because even the fanciest LIDAR maps still cannot detect exactly which insect is feeding on what type of leaf, or which epiphyte shows yellowing in what layer of the crown. In short, overflights still cannot substitute for the observational powers of an experienced arbornaut (or even an inexperienced arbornaut who happens to be a crackerjack botanist).

Another important result of our conference in India was the publication of a symposium volume. Such productions are important for two reasons: first, they give emerging scientists the opportunity to publish alongside professionals from regions with more developed scientific programs, elevating their in-country status so as to have a better chance for promotions and local recognition; and second, such publications disseminate the newest findings. A female forester from Cameroon was thrilled to publish in the proceedings, and it undoubtedly led to her promotion back home. One of the first chapters, written by two young biologists, Alex Racelis and James Barsimantov, asked a fundamental question: Does more investment in tropical research within a specific region contribute to lower deforestation? To answer this, they correlated numbers of scientific papers (a metric of scientists and their research dollars invested) with conservation effectiveness, and they found higher publication output did not correlate to saving more forests. Our current scientific process may not be leading to successful forest conservation; we scientists should not remain content to simply publish results in technical journals as a primary accomplishment. The India conference reminded me about the importance of measuring my own success not by peer-reviewed publications alone, but by international students mentored, community engagement in countries where forests lacked research infrastructure, and acreage of tree canopy conserved.

Forest conservation in India was at a crossroads. The country had exploded with technology and modernization almost overnight. While canopy biologists were gathered at a conference, India's major automobile company, Tata Motors, launched a global sensation, its new Nano, a tiny 33-horsepower car costing a mere $2,500. India has a growing middle class—but its natural resources, especially trees, were becoming more and more squeezed. In the early twenty-first century, native forests in India were declining by 1.5 percent to 2.7 percent per year, giving way to tea and exotic timber plantations. Conservation biologists call this cryptic deforestation, because it is difficult to calculate how much old-growth or primary forests is increasingly replaced by crops, using an aerial survey of green cover, because the green canopies cannot be differentiated. How India (along with China) ramps up her consumption of natural resources and whether billions of people in those two countries adopt sustainable practices will largely determine the fate of our planet in the next few decades. In many ways, it is not fair. Western nations used their consumerism ruthlessly over the past few decades, but now the world recognizes that this rate of consumption is not possible for eight billion people without dire consequences.

After the conference, T Ganesh, Soubadra, and I taught a climbing workshop for future arbornauts at a nearby reserve called Honey Valley. The word "nearby" is relative. What looks awfully close on a map, even in terms of mileage, can take hours of travel—due to road conditions, livestock, congestion, trucks, oxcarts, or weather. Our destination was one of the closest primary evergreen rain forests, located at 4,100 feet in elevation with a prominent mountain peak called Thadiyandamol as a popular hiking destination. This region, just over 125 miles from Bangalore, derived its name as one of the largest honey-producing apiaries in India, but an epidemic of Thai sacbrood virus disease decimated the honeybees and eliminated beekeeping for the local residents. Instead, they quickly turned to ecotourism, creating lodes for birdwatching and rejuvenation, a wonderful illustration of the nimble versatility of the Indian spirit. Soubadra and T Ganesh had booked several cabins for students and instructors, providing their Bangalore students with a unique opportunity to train as arbornauts with a few international conference experts. Soubadra and T Ganesh organized a training session in climbing trees with single-rope techniques (SRT), as well as some lectures about canopy research. Since conference participants were already in India for the event, it was not too much effort to add three more days onto our agenda and do our part to train India's next generation of canopy scientists.

Our day-long drive started out with a dawn breakfast about one hour outside of Bangalore, and then a good five hours of driving to this "nearby" site. We feasted quickly on a few shared dishes with amazing flavors and sauces. I was always a bit hungry in remote field locations, usually eating cautiously because my Western microbiome sometimes rejected the variety of extraordinary spices composing Indian cuisine; I loved every bite, especially the smells. Upon arriving at Honey Valley, we marveled at the stands of tall rain forest trees, mostly vedippala with its keystone fruits that seem to feed almost every food chain in the forests of the Western Ghats. I was assigned a small room in a shared cabin, furnished simply with a single bed, side table, lamp, and one chair, with a shared bathroom down the hall. Looking out my window, I was appreciative of our transformation from chaotic Bangalore streets to a tranquil green oasis in the mountains. It was great to envision an increasing number of Indian families taking sanctuary in nature as ecotourism took hold in India. It was also obvious that many acres of Honey Valley were already cleared, farmed, and no longer served as primary forest ecosystems, so the remaining pockets of forest had an elevated value for nature conservation. After we quickly settled into our cabins, T Ganesh surprised us with a wildlife demonstration. He had invited a herpetologist, who not only conducted snake research but also worked part-time as a cobra whisperer, someone who detects and extracts these venomous snakes from farm cottages. The incidence of cobras seeking shelter in rural houses was commonplace and dangerous, creating a huge source of fear for farmers. What a livelihood! This young man brought several containers with him. An enormous and aggressive cobra exploded out of the first box when he cracked the lid. Using a snake stick, he deftly controlled the slithering beast. We all gasped and took several steps backward. I snapped some photos but was relieved to remain in the back row for this display. Maybe I am slightly ophidiophobic despite my many years of snake encounters?

Next—climbing classes! Our student group was composed of eight women and twelve men, all interning at ATREE in Bangalore. To climb trees in traditional dress was not easy, and I made a note to myself to ask if they would approve of my bringing a suitcase full of khaki pants for women the next time. My fellow arbornaut Tim Kovar acted as our lead technical instructor, and I served as the science lecturer. Tim, an inspirational teacher with a perfect safety record, has climbed with me all over the globe. We have had some amazing experiences, ranging from teaching an eighty-year-old woman whose lifelong dream was to climb a kapok tree in the Amazon, to hosting a hundred children in North Carolina for a public tree-climbing day, to training my wheelchair-bound students in single-rope techniques. Both of us were committed to "tree climbing for all." During our cobra session, Tim quickly set up a climbing site in the nearby rain forest and rigged three trees with ropes dangling ninety feet overhead. The Indian students all traipsed out into the trees, cautiously stepping around dense vegetation after our intimate introduction to cobras, and sat down in a small clearing with a perfect view of a few vedippala canopies affixed with climbing ropes. Unlike Australia, we did not have hundreds of leeches crawling up our legs, and different from Central America, we did not have chiggers giving us multiple bites along our underwear lines. In fact, the Indian rain forest was relatively benign in terms of dangerous beasts, except for the odd cobra and a few large creatures such as elephants and tigers. Despite those differences, the Indian rain forest reminded me of Australia—many similar tree genera, a similar structure in terms of vine and seedling density on the forest floor, and similar levels of foliage and canopy height. Structurally, the forests of India and Australia reflect a common heritage, as do their evolutionary overlap of similar plant families. But India's forests contain species found nowhere else in the world. The Western Ghats, recognized as a global biodiversity hotpot, contains 5,640 flowering plants (of which over 100 are endemic), 165 freshwater fishes, 76 amphibians, 177 reptiles, 454 birds, and 187 mammals; all of these numbers increase as new discoveries unfold.

By the end of one day, we had converted twenty frightened Indian biology majors (almost half female) into potential arboreal explorers. Training the next generation is critical in an emerging field like canopy ecology, an important metric of scientific success. Soubadra and T Ganesh were wise to capitalize on the presence of international scientists at their conference and offer a field workshop as a brilliant way to give their students a life-changing opportunity. I laugh to recall myself struggling up their rickety ladders in vedippala branches three years earlier; I had brought them gifts of ropes and climbing gear a year later. If we could integrate field research with India's community values and religious sanctity, my Indian colleagues and I would achieve success through local, aka bottom-up, community-based conservation. Through an arbornaut's lens, I have watched India evolve a whole-forest approach to field biology, using new tools like camera traps to document biodiversity and single-rope techniques, leveraging the value of sacred trees and determining the importance of canopy cover for animals like tigers or leopards on the forest floor as well as primates up in the treetops. When arbornauts collaborate in countries like India, we can import best practices to jump-start their scientific advancement, and hopefully stay ahead of threats to the integrity of their ecosystems. By adopting new tools and working with international scientists, Soubadra and T Ganesh have achieved major discoveries in the branches above their heads.

T Ganesh pioneered the use of camera traps in India's forest canopies, starting with the upper branches of *Cullenia*, where he documented plant-animal interactions. Visitors to the *Cullenia* flowers captured by his lens over a three-year period included: lion-tailed macaque (*Macaca silenus*), Nilgiri langur (*Semnopithecus johnii*), giant squirrel (*Ratufa indica*), giant flying squirrel (*Petaurista petaurista*), Malabar spiny dormouse (*Platacanthomys lasiurus*), dusky striped squirrel (*Funambulus sublineatus*), and brown palm civet (*Paradoxurus jerdoni*), plus two types of bats (*Cynopterus sphinx* and *C. brachyotis*), but only near disturbed edges. In addition, sixteen species of birds, two types of bees, and a few species of butterflies, ants, and moths were recorded. Lion-tailed macaques also ate the trees' leaves, buds, and flowers, as well as its fruits. In the Western Ghats, *Cullenia* flowered during the dry season from December to April with three hundred to thirty thousand brownish-yellow tubular flowers arranged in dense cauliflorous clusters. The succulent fleshy base of the flowers is soaked with nectar, so animals seek this part, discarding the remaining floral parts in the process. No herbivory studies have yet been undertaken for vedippala, but I hope to return someday and tackle this exciting fieldwork. Meanwhile, Soubadra has documented pollination of many Indian trees by flying insects (especially butterflies), although she admits few observations exist for Indian birds or arboreal mammals as compared to more extensive research in other tropical regions. In the Western Ghats, only two of eighty-nine tree species are pollinated by mammals; trees of Australia, Africa, and the tropical Americas exist in stark contrast, with many important animal pollinators including marsupials, rodents, giraffes, and even primates. Why does India appear to have more insect pollinators than mammals or birds? No one knows, but with their arbornaut toolkits, Soubadra and T Ganesh just may solve this mystery.

One of the toughest challenges in my career as a global arbornaut has been the transition between the rigors of fieldwork and a return to daily family life. No matter how many times I have moved between the two, the emotional upheavals and physical toll of global travel and working in unfamiliar places are never easy. This was especially true in India. When Soubadra, T Ganesh, and I headed into the field to rig trees or survey biodiversity, we frequently transitioned through several economic and cultural worlds. Consider this journal excerpt describing one day in a tiger reserve near the Western Ghats mountain ranges, plus the overwhelming reentry into the "urban jungle":

6:00 a.m. Alarm clock rings. Still dark, a dawn chorus of hornbills announces sunrise in Nagarahole National Park near the remote village of Karapura. Struggling from a cocoon of thin blankets on a rock-hard wooden cot at the Kabini River Lodge, I fumble for my khakis. The air is chilly but exhilarating. We are here on a mission—to survey big cats and discuss their conservation in the context of forest habitat. The Western Ghats of southwest India is one of an estimated twenty-five global biodiversity hot spots, as defined by the NGO Conservation International.

6:15 a.m. Two Indian waiters dressed in flowing white trousers, traditional dhotis, knock on the door, bringing tea as a morning wake-up. Ants swarm on the doorstep, enjoying the crumbs dropped from yesterday's afternoon tea. I dance to shake them off my bare feet.

6:30 a.m. Still shivering in darkness, our team of four Indians and one American jump into an open-air jeep. Along the track, women are already awake and hauling water in plastic jugs from the village well to their thatched huts. Wood-burning cookstoves emit small spirals of fragrant smoke into the sky, announcing breakfast. As a result of the cook fires, black soot insidiously enters the air, creating health issues for women who stir the pots and inhale particulates, and that same aerial dust accelerates ice melt in Himalayan glaciers several hundred miles away by darkened ice surfaces melting faster.

6:50 a.m. Our noisy vehicle passes under a mango canopy where two guards peer down from a rickety treehouse. In their aerie, watchmen guard the village crops from marauding tigers and elephants.

7:00 a.m. At the park gate, armed rangers inspect our paperwork. Entry permits are issued in limited numbers. Despite best efforts, poaching is rampant in India and threatens big cats with extinction.

7:00–10:00 a.m. Still shivering, we bump along a rough track in our open-air vehicle with all five senses on overdrive. Our guide listens for a specific bird call that signals danger, providing a clue to the location of a predator (most likely: big cat).

8:13 a.m. The driver screeches to a halt, whispering, "Listen." A female Sambar deer shrieks nearby. In the branches above our dirt track sprawls a leopard, chewing its fresh kill of a fawn. The laws of the jungle are tough, and one hesitant step by a mother and her offspring resulted in a casualty. This dominant predator retreated to the treetops to consume breakfast. We all marveled at the notion of a ground-based predator using the canopy as a safe haven, making our whole-forest approach to tree ecology all the more relevant. Leopards, like tigers, are rapidly declining throughout India, and both need healthy forests to survive.

9:15 a.m. After seeing two more leopards but, alas, no tigers, we exit from the forest and prepare for reentry into the urban landscape. Due to its permit restrictions, the park would not allow us to get out of our four-wheel-drive vehicle, so we were limited to observations, photographs, and note-taking as our means of "studying" forest canopies in India's tiger reserves. This visit was more like a quick ground-truthing exercise, not a full-blown research program, which might entail several years of similar observation days. But again, as Soubadra, T Ganesh, and I ramp up our international collaborations, it is important for us to share sights and sounds of forest ecosystems, giving rise to conversations that lead to new ideas. The entire drive was full of animated discussion. To travel home after this quick visit, I will make a huge transition from a remote wildlife reserve through four airports and several dense cities, ultimately disembarking in Florida. No time for a shower, I pack in two minutes and quickly gulp a few mouthfuls of traditional vegetarian fare.

10:10 a.m. On the road, I am bound for the local airport as the first leg of a long journey. Indians think nothing of a six-hour taxi drive, dodging goatherds, oxcarts, children walking to school, bicycles groaning under enormous loads, roaming cows, and dilapidated public buses. I ration my single water bottle, as any explorer always does. My mind struggles with the abrupt transition from oxcarts to jumbo jets over a half-day time frame.

4:45 p.m. The domestic flight from Bangalore to Mumbai is delayed. I frantically stand in a long line to find another flight. Luggage is nowhere to be seen.

8:30 p.m. Late arrival to Mumbai's domestic airport leaves a meager two hours to connect at the international terminal approximately twelve miles away. Miraculously, my suitcase appears on the belt. The wait for a taxi is over an hour. Against better judgment, I squeeze aboard the interterminal bus at 9:15; I have just over one hour before the New York flight departs.

9:35 p.m. At the international terminal, passengers stampede through a narrow gate, frantically waving passports and boarding passes. Suitcases are flying. The air is rich with expletives. Chaos reigns.

9:45 p.m. At the United Airlines check-in, a ticket agent whisks me through. Will I make it? Dripping with sweat, I look feverish, so a health officer stops me to check for swine flu. No, I reply, just perspiration from flight connections. Running, I am the last to board. Sinking into a plush seat is sheer ecstasy. This sixteen-hour flight with meals on trays, soft seats, toilet paper, and headphones feels like a luxury instead of a curse. As an arbornaut who often experiences the transition from trees to cement, I muse at the world's definition of "civilization." With just over 50 percent of humans now inhabiting cities and appreciating the cultural and physical amenities provided by urban life, field biologists remain a minority who would rather confront leopards in a green jungle than traffic in concrete ones. As a bustling, emerging country, India's trees face degradation in the wake of "progress." I hope our canopy research efforts will generate enthusiasm for forest research and conservation in India.

Many emotions overwhelm me not only during global travel but also when I'm climbing, especially in a remote place far from my children. Although ascending into the treetops is a joyful experience, it is also fraught with anxiety, caution, and a healthy dose of fear. Climbing trees requires a large amount of trust—in the strength of the branches, the cooperation of Mother Nature, and the grip or "dirt" (the person who stands on the ground to monitor safety). In countries like India, Ethiopia, and Peru where I undertook many years of research, loneliness pervaded at times. It was not always easy to communicate from the upper branches to the people standing below due to language barriers. And it was nearly impossible to communicate with my family half a world away due to lack of connectivity. I often missed holidays, including birthdays, and couldn't easily convey a sense of the remoteness of many field sites to my family. The demands of international fieldwork make it tough to find the energy for friends and family upon returning home after a rugged expedition. While I always had good intentions of sharing arboreal tales with both boys upon arriving home, I was usually fighting jet lag and couldn't wait to brush my teeth over a real sink and collapse in a soft bed. All jobs have trials and tribulations—but life as an international arbornaut certainly has its own ups and downs, literally and figuratively!

One of the ups comes from the fact that when I'm dangling from the end of a rope, wildlife usually accepts me as a fellow treetop denizen. I've spent many solo hours in Massachusetts temperate forests watching a sapsucker devour caterpillars, sharing eucalypt branches with koalas in the Australian outback, or admiring ants hauling bits of foliage much larger than themselves in the Amazon. India was no exception. At KMTR, working with Soubadra and T Ganesh to implement the canopy toolkit to survey India's biodiversity, I had some close encounters with our primate relatives. One major project involved establishing permanent plots and monitoring the seasonality of fruiting in the canopy dominant, vedippala. With greatest trepidation, we climbed some precarious twig ladders affixed to several high branches, allowing permanent access, and sat amid a troop of our distant relatives who were enjoying a fruit feast. Like tigers, lion-tailed macaques depend upon vedippala for habitat, but these primates also feed on the fruits and live and play in the branches, whereas tigers are limited in range to the forest floor. Soubadra risked life and limb to study these enormous crowns, bravely using her hand-hewn, rickety ladders (which reminded me of my roughshod scaffolds in the Scottish Highlands). Meeting macaques was thrilling, and the forest birds of India greatly expanded my vocabulary with amazing names like scarlet minivet, brown-cheeked fulvetta, plain flowerpecker, and little spiderhunter. Most were insectivores, and despite their palate for bugs, depended for survival on vedippala. Insects are attracted to eat the tree's fruits; birds and reptiles in turn consume the insects, and monkeys join the lineup to eat insects, leaves, and fruits. Ecosystems are complex chains of who eats whom, and India's tropical forests are no exception. The health of this country's canopy in turn supports iconic wildlife such as tigers and leopards, reinforcing the interconnectedness of ecosystems from top to bottom.

Cullenia exarillata is an example of what ecologists call a keystone species, meaning it impacts the health of an entire ecosystem. Another example is the starfish along North America's Pacific coast, whose predatory habits prevent any single species from dominating and outcompeting others in intertidal pools. Similarly, alligators represent keystone species in Florida, and crocodiles in Africa, making deep water holes that form oases for other wildlife during droughts. In India, elephants are considered keystone species by acting as ecosystem engineers, sometimes uprooting trees, which in turn has a cascading impact on other herbivores because it destroys the foliage supply, reduces canopy cover, and leads to new tree species growing up in the open space. Because tropical forests have not been as extensively studied as many temperate ecosystems, fewer keystone species have been identified there. Field biologists still don't really even know what lives in the tropical forests. The notion of a keystone role remains controversial among ecologists because it is not easy to tease apart the complex relationships in ecosystems. But if a species is labeled as keystone, it usually becomes a primary focus for conservation, because if its presence declines, the landscape is predicted to alter or degrade.

Although technology can transmit information around the globe in seconds and photograph outer space, many mysteries closer to home remain unsolved. Scientists have not yet figured out the ecology of vedippala seed rain, germination, and pollination strategies, nor any of the complex interactions of India's biodiversity within the crown of this keystone tree species. A dwindling number of tigers and macaques surely hope we can do better. Their survival not only depends on the continued fruiting of this tree, but also on a better understanding of how the whole forest works. A recent headline in the *Bangalore Mirror* read, "Tiger Spotted in Sahyadri Reserve." One tiger sighting makes national headlines. In this case, biologists had not seen the animal but genetically analyzed its scat on a trail in the reserve. Less charismatic, but equally important, will be the follow-up research on the canopy under which that tiger lives, and how to keep it healthy. Will Princeton University be satisfied with an extinct mascot? Can circuses survive without tigers or elephants in the big top? The world would never be the same if some of India's most beloved wildlife became extinct, and the next few years will likely determine the fate of these noble creatures.

Another large and charismatic Indian animal is equally at risk, the Indian or one-horned rhinoceros (*Rhinoceros unicornis*). I had the honor of riding an elephant to count rhinos in Kaziranga National Park with Soubadra and T Ganesh during another field visit. These amazing animals numbered only 2,093 individuals throughout the entire world at the time, depleted by poachers for the Chinese medicinal trade. I found out why before we'd even begun the counting, while sitting at a hotel during dinner and listening to an American biologist brag about watching rhinos copulate for twenty-five minutes, which he claimed was a world record (for rhinos, that is). When this stamina is broadcast, whether it is true or not, people associate rhinos, particularly their horns, with sexual prowess. The day of the count, we arose in the dark to reach the park gates before dawn and find our elephants hired for transport. En route we stopped at a street-side kerosene burner where its owner was selling the tiniest cups of tea I had ever seen, probably one medium-sized mouthful, steeped with milk and sugar. Fortified, we headed into a dark fog, but soon found the meet-up site with saddled elephants, all milling around near a ladder stand built for us to climb up and board our creatures. Soubadra and I rode on Rakumana (named from Indian mythology). Kaziranga National Park boasted the world's largest remaining population of rhinos, as well as another possible world record of ninety tigers. Riding on native (but tamed) Asian elephants, we were not recognized as humans and so got within twenty feet of the rhinos! It was incredible to view a mother and her baby foraging in the tall grass, shrouded in morning fog. I glanced past our elephant's butt to catch a streak of orange and black zipping through the underbrush. I can't prove it was a tiger, but the anxious behavior of nearby hog deer suggests it was, so I am sticking to my story. By sunrise, I felt spiritually connected to this special place, having seen some of the world's most unique and threatened species from the vantage point of an elephant's back. As we left, local rangers were on strike. It was a conservation conundrum. Because a recent survey had declared that this park housed a high density of tigers, international NGOs wanted to make it a dedicated cat reserve. But if this status were bestowed, the park would also be subject to more limited visitation, putting local people out of work; some might be forced to relocate. The local rangers wanted to retain the park's current designation as a rhino reserve and have nothing to do with tigers in any official capacity. If the locals lost their jobs, some could resort to poaching. It was easy to understand why indigenous communities became resentful of international conservation groups who lived half a world away and didn't always understand the local issues. Because of this conflict, my colleagues hid me in the back of our jeep as we departed, so Indian rangers would not suspect I was a tiger advocate. This illustrated the classic interference of well-meaning global stakeholders from the viewpoint of locals trying to juggle livelihood and conservation.

Without question, the majority of Indian citizens now live in urban environments. The fact that India has successfully conserved 21 percent of her primary forests is a tribute to the country's respect for trees, but it's also due to massive urban migration. As India's economy and population has grown, the pressure on natural resource exploitation has increased. This has prompted the consideration of unconventional partnerships to promote conservation. In his controversial book *Half-Earth*, the biologist E. O. Wilson advocated for conserving half of the planet for one species (humans) and the other half for the other 99.9 percent of species. His list of a few "best places to save" included India's Western Ghats, and initial canopy research there confirms the uniqueness of the region's biodiversity. The Western Ghats exemplify India's strong sense of spirituality with regard to trees. Indigenous people predominantly practice Hinduism, where the Bhagavad Gita preaches an important moral obligation to worship nature. India houses the highest concentration of sacred forests in the world, with approximately 100,000 to 150,000 stands. Historically, Indians worshiped icons of nature, and even today, Hindu families light a lamp under a sacred tree as a spiritual gesture. Most temples and villages feature at least one large fig (*Ficus religiosa*), also called peepal or Bo or bodhi tree. *F. religiosa* was the religious species assumed by Lord Krishna, according to the Bhagavad Gita. This notion of sacred trees may have stopped British rulers from felling India's primary forests, because they so greatly respected the indigenous religious beliefs. Similarly, the king cobra (*Ophiophagus hannah*) was a deity associated with many sacred groves in lower-caste communities, giving people an incentive not to kill it. Such interactions of religion and nature illustrate how sacred beliefs directly aid the conservation of a threatened species.

India's natural resources are inextricably linked to religion, and successfully so. Today, many sacred forests are managed by joint groups of Hindu families or temple trusts, and over five thousand fragments exist in the Indian state of Kerala alone, at the southern range of the Western Ghats. Integration of sacred groves into state-owned protected areas has become a viable solution to ensure continued protection in the twentieth century. For example, Kerala's government offered financial support to fence and protect over five thousand sacred groves throughout their state. Many trees in India remain protected today because of this unique religious sanctity, instead of their timber value as is the case in many Western countries. It continues to challenge forest scientists who seek more creative metrics to value forests in Southeastern Asia, in the context of religion as an overarching conservation mantra. Perhaps the number of prayers delivered in a sacred forest could somehow be translated into dollars and cents?

As the world learns how to place on trees values other than timber, India has ramped up her forest restoration activities to offset carbon dioxide emissions. During 2019, the state of Uttar Pradesh in India hosted a twenty-four-hour tree-planting blitz, at which time the citizens planted over 220 million seedlings (at least one per person in this populated region) in one day, practically a world record in terms of quantity versus time. Thanks in part to the exact data collected from canopy research about carbon dioxide intake and ultimate carbon storage of whole trees, forests are now recognized as an important weapon against climate change. It is estimated that a billion trees can remove 25 percent of our current carbon dioxide emissions, and India already planted a quarter of that total in one day. Although old-growth stands are the most effective carbon storage units because their large trunks and crowns store more carbon, those seedlings planted in Uttar Pradesh will someday become adults. This additional focus on tree planting is a wise investment for India's future, and essential for a country of over one billion people.

⫸ Vedippala ⫷
(*Cullenia exarillata*)

A SCOTTISH BOTANIST NAMED ROBERT WIGHT defined the genus *Cullenia* from one sample collected near Coimbatore, Madras, India, and he differentiated it from the genus *Durio*, its closest relative, by observing that the lobes of the staminal tube were much longer. These species are members of the family Bombacaceae, a global distribution of tropical trees including the silk floss, great kapok, and java cotton. In botany, the first described individual of a new species is called a type specimen. This sets the standard for identification, and type specimens are safely accessioned in herbaria at museums. In a recent publication, *Cullenia* was revised to reflect updated revisions to the classification of this genus, after careful analysis of morphological variation. To illustrate the technical and almost undecipherable jargon of these taxonomic descriptions, here is the revised botanical text:

Trees, tall, evergreen, the branchlets lepidote. *Leaves* alternate, peti-
olate, the stipules fugacious, the blade simple, entire, the upper surface glabrous, the lower surface covered with numerous overlapping
scarious-hyaline peltate scales. *Flowers* ramiflorous, densely fascicu-
late on protuberances of the old wood, pedicellate, the pedicels artic-
ulate near the middle, densely lepidote; epicalyx valvate, irregularly
3–4-lobed at the apex, splitting to the base on one side and deciduous, densely lepidote without; calyx valvate, more or less irregularly
5-dentate apically, carnose, deciduous, densely lepidote without, the
scales peltate, scarious-hyaline and overlapping; corolla absent; staminal tube 5-dentate-lobulate to 5-lobate, each lobe with 7–11 stamens
along the margins; stamens with short filaments, the connectives irregularly spherical to club-shaped, covered with numerous stipitate
1-locular thecae, the latter more or less globular, densely covered with
minute deciduous mammiform processes, dehiscent by means of an
annular transverse median slit (circumscissile).
—ANDRÉ ROBYNS, "Revision of the Genus *Cullenia* Wight
(Bombacaceae—Durioneae)," *Bulletin du Jardin botanique National
de Belgique* 40, no. 3 (1970): 241–54

When I first started botany training, I could not survive without a
botanical dictionary, so that each word could be translated into everyday language. But with practice, all those ovaries, pistils, and stamens
took on extremely precise meaning, thanks to many centuries of botanists who formulated taxonomic nomenclature. For field biologists,
the challenges of finding species in nature is enormous, but equal to
that is the daunting task of classifying all those discoveries. It can take
a lifetime to figure out how to explore ecosystems and find species, but
another lifetime to learn their classification. The taxonomy of plants,
or any other biodiversity for that matter, is a moving target. As better
microscopes and improved DNA analyses become available, the evolutionary tree of organisms is revised, updated, and sometimes found
to be plain wrong. The genus *Cullenia* seems stable with three species,
of which *C. exarillata* serves a keystone role within India's forests.
Although the taxonomy of vedippala is known, its ecology and

life history are not. This iconic species is not only a keystone but appears to exhibit critically important mutualisms with its pollinators, including bats, birds, and several mammals. Again, not enough research exists about India's canopies to fully understand the complex interactions of her forest ecosystems. In addition to providing homes to many animals, *Cullenia* harbors about 40 percent of the country's recorded epiphytes, so its crown is a major hub of floral biodiversity. The edible fleshy sepals of its fruits provide a meal for some of the larger pollinators, including sloth bears and civets. Other arboreal creatures eat the fruits, such as Nilgiri langurs, lion-tailed macaques, Hanuman langurs, Malabar giant squirrels, three-striped palm squirrels, and terrestrial rodents. Some fruits fall to the ground or into streams, where another gang of predators awaits—insects, mollusks, crabs, fish, rodents, squirrels, and some ground birds. Other iconic species rely on vedippala overstory for their understory habitat, including tigers, leopards, and Asian elephants. There is a scenario dubbed "empty forests" whereby clearing and fragmentation create isolated stands of trees devoid of biodiversity, because the insects and birds are not easily able to repopulate, which requires a long-term recovery process. To continue to serve as India's Noah's ark by providing a refuge for so many species, vedippala needs to remain a focal tree in healthy stands. T Ganesh and Soubadra, the dynamic duo who lead important field research on India's trees, represent the human version of "type specimens" as a new breed of conservationists who operate without large budgets, significant government funding, or nonprofit underwriting. Such small teams, working tirelessly in countries where research dollars for forests remain sparse, are the lifeblood of current global conservation. Similar individuals exist in Sri Lanka, Cameroon, Ethiopia, Bhutan, and many other biodiversity hot spots. They are the understated heroes of our planet, and I am dedicated to helping them advance their efforts by providing tools, ideas, support, and international collaboration.

9

A TREETOP BIOBLITZ

⧼⧽

Counting 1,659 Species in Malaysia's Tropical Forests in Ten Days

T HE HILLSIDE WAS BLANKETED in a dense cotton ball of fog, mystically enshrouding all the trees. Straining to see any upper limbs, the tree rigger Tim Kovar, whom I had dragged around the globe for years as my arbornaut partner, aimed his Big Boy slingshot and fired. Up, up it went and over a sturdy branch of dark red meranti (*Shorea curtisii*). Our first canopy tree of the renowned Southeast Asian family Dipterocarpaceae was rigged! The first climb was almost spiritual as we ascended into the white blanket enveloping Penang Hill. A troop of macaque monkeys chattered below us, perhaps because we had invaded their home? On another day, a small band of dusky leaf monkeys blessed our climb with their presence and an elusive colugo gazed quietly from an adjacent perch. The Malaysian forest, with its crowns stretching up into the heavens (mostly dipterocarps), is brimming with life, yet many of its uppermost secrets remain undetected as part of the unexplored eighth continent.

So how did I get to Southeast Asia to work on dipterocarp conservation? If you google "canopy biology," my name comes up near the top of the list. At least that's how a canopy construction company in Canada located me. The company was in a bind: a client requested a scientist to critique the placement of a new walkway under construction. Would I fly to Penang, Malaysia, to make a scientific assessment of their proposed site? In 1996, I had taught climbing classes at the Bogor Botanical Gardens in Indonesia, awestruck at the time by two local colleagues, Sofi and Mati, who managed to climb in their traditional head coverings plus gorgeous flowing-yet-impractical-for-climbing garb. In 2013, I was invited by the royal family of Johor, Malaysia, as the keynote speaker for the country's Conservation Day. Even longer ago, I had stopped in Kuala Lumpur en route to Sydney University on the flight that launched my graduate adventures. I was thrilled at the opportunity to return to the world's tallest trees, and even happier to provide scientific oversight of a new walkway.

A few months later, my entire world was transformed, with a new canopy walkway project in Penang and the challenge to execute a biodiversity survey of their local tropical forest within a ridiculously short time frame to meet a deadline for a UNESCO World Heritage nomination of the site. The success of this conservation and ecotourism project was due to the vision of one local philanthropic business entity, the Cockrell family, who operated The Habitat Foundation and underwrote the Penang walkway thanks to their passion for local forests. This was my first experience pitching big conservation projects to a business entity, where one major donation can jump-start a project faster than the conventional, albeit inefficient, process of writing grants, rewriting grants, then invariably reducing the budget and resubmitting the grant again. Continuing on my mid-career trajectory of sharing the canopy toolkit with different countries, I was focused on becoming a better conservation steward, not just a conventional tree scientist. In India, I first learned the importance of sharing the arbornaut toolkit with other countries; in the Amazon, I leaned on the contributions of citizen scientists to achieve greater outcomes; and in Ethiopia, I learned to measure success by acreage of forest saved, not just amassing technical data for academic publications. In Malaysia, in addition to partnering with a business entity, not conventional academic funding, I also incorporated a new public-science-based activity, the BioBlitz, to survey species.

BioBlitzes, defined as rapid species counts involving a crowdsourced activity over a short span of time, started in 2006 when some children were searching for nine-spotted ladybugs in Washington, DC. Approximately five hundred species of ladybugs exist in North America, each with different numbers of spots and colors. For most of the twentieth century, the nine-spotted ladybug was the most common; it was even declared the state insect of New York. But the invasive European seven-spotted lady beetle quickly and aggressively began replacing the native species. The Lost Ladybug Project was initiated to document any sightings of the endangered nine-spotted ladybugs during a summer season. Answering the challenge, two children, Jilene and Jonathan Penhale, found a nine-spotted ladybug near their home in Northern Virginia. Hundreds of volunteers turned out to look for additional specimens, and their search was considered to be the first BioBlitz, whereby many volunteers search a specific area over a defined time frame. Unfortunately, no additional specimens were found at that location, although subsequent surveys found viable populations on Long Island. Online resources allowed participants to amass over forty various images of ladybugs with nine spots around the country. Today, similar surveys sometimes yield thousands of images, but the nine-spotted ladybug was an important pilot.

Since then, BioBlitzes have become a great tool for amateur naturalists and professional field biologists to team up and find, identify, and survey species during a short burst of time. When images of species, along with their identifications, are uploaded on social media, the value of these surveys is enhanced by making the results available to a global audience. This concept played right into my passion for public science communication and the challenges of finding species in complex forest habitats. The notion of engaging citizen scientists was a game changer. However, a recent scientific review of public engagement in the journal *Frontiers in Ecology and the Environment* stated that BioBlitzes remain more common in industrialized countries, mainly because people have more volunteer time and easier access to technology for online data collection. So it felt all the more important to work toward greater inclusivity in the sciences by establishing public engagement efforts outside of Europe and North America.

In Penang, I met my client—a family with local roots who also operated a diverse international company. The developers and funders of the proposed aerial trail were one and the same. They wanted to give back to their community by building an eco-park that would educate the public about tropical rain forests and believed a canopy focus made good economic sense in terms of attracting visitors by providing a unique rain forest experience. This was the first conservation project I had worked on for which someone wrote one big check; my role was to ensure that the science and conservation outcomes would be effective and long-lasting. The first assignment was to inspect the walkway site and design education interpretation along its path. Second, when the Canadian construction team suffered some glitches in their construction process, I offered insights based on prior walkway construction experience. They built not one, but two walkways, using a special design element called a ribbon bridge, at a cost of 8 million Malaysian ringgits (RM), the equivalent of $1.8 million. This type of skywalk utilizes cement for its underpinning, meaning the bridges can literally hold hundreds of people at once. It may have been overkill, but it resulted in two elegant bridges, spanning over 450 feet and rising over 125 feet high, with a large platform in between. I hosted several classes and walks with the park's education staff, providing them with stories and facts about the eighth continent relevant for their daily role as interpretive guides. And third, the funders wanted to know how to turn their walkways into a world-class canopy research site; I suggested three things: (1) conduct a biodiversity survey to showcase the primary forest by bringing in an international team to partner with local scientists; (2) secure the landscape with permanent conservation status; and (3) create a field station offering treetop access that would attract biologists to conduct long-term research. I had overseen one or two of these three components at other field sites, but never had all three so conveniently packaged into one project. Suddenly, I found myself co-organizing a biodiversity survey, collaborating on a UNESCO World Heritage site nomination, and advising on a future field station.

Ever since Charles Darwin collected thousands of species in the tropics during his expedition on the *Beagle*, biodiversity has become synonymous with ecosystem health and resilience. But the toolkit for tropical species surveys has not advanced much over recent decades as compared to genetics, agriculture, or even astronomy. Conservation International developed their Rapid Assessment Program (RAP) in the twentieth century, whereby they engaged teams of experts to conduct surveys of diverse global ecosystems as a means of determining which habitats were the most important for saving. In the twenty-first century, the notion of citizen science is expanding the scope and breadth of biodiversity surveys using rapid assessment, now called a BioBlitz when the public is engaged. Using volunteers over a finite period of time, surveys of urban parks, beaches, or almost any ecosystem can be executed rapidly to assist with conservation management, or to assess endangered species. In addition to engaging the public, cell phone apps such as iNaturalist (www.inaturalist.org) have allowed integration of species data into social media platforms that also serve as a library for the scientific community and conservation practitioners. When I was the chief of science at a California museum, I had the good fortune to fund and engage iNaturalist as an arm of our museum research team, which ultimately spotlighted it within academic and museum circles. It is now comanaged by National Geographic, whose international brand has enabled the app to expand its global reach.

The Habitat eco-park, situated on Penang Hill in the state of Pulau Pinang, Malaysia, attracts visitors from the city with its nature trail, two ribbon bridges (called Langur Way), and another aerial trail consisting of a series of zip lines. Collectively, the Habitat funders, local university scientists, and government officials aspired to submit a UNESCO nomination for the adjacent 7,285 hectares of forest, along with 5,196 hectares of offshore marine sanctuary. By definition, biosphere reserves are areas of marine or terrestrial ecosystems that receive international recognition within the framework of UNESCO's Man and the Biosphere Programme (MAB). They are established to link relationships between humans and nature, so both biological and cultural resources are required for a successful site. This region was characterized by primary lowland tropical rain forest, dominated by the trees of the family Dipterocarpaceae. It was an exciting proposition to find out what biodiversity inhabited these hillsides and recruit both local and international experts to comb the hillsides from the top to the bottom of the dipterocarps.

Despite easy access to a major city with a distinguished university, no organized natural history survey had been conducted on Penang Hill. During the first visit to meet with the funders and see their site, I also set up local meetings with Universiti Sains Malaysia (USM), the country's major science institution, and gave a talk about the canopy, highlighting their incredible backyard biodiversity in Malaysia. The biology department chair invited me soon after to become a research professor, an unpaid but respected title. Since that first seminar, I worked closely with USM and as a result of their enthusiasm, we created a working group to plan a BioBlitz, the first part of a three-stage plan to make Penang Hill an international research center. The corporate donors generously awarded me a grant to invite participants and, with careful administration, funded approximately thirty international scientists and another fifty USM biologists and students, plus a few government officials and staff from the eco-park. We carefully matched up local scientists with international experts so the surveys would encourage future collaboration. I wrote an additional grant to fund a team of teenagers who livestreamed a virtual field expedition to schools around the world, so in total, the expedition included 117 participants. Of course, we did not manage to avoid logistical drama. A few scientists wanted more funds than the budget allowed, some wanted to bring their children (which was fine but required extra homework to ensure their safety during fieldwork), and others wanted to take a vacation after they completed the blitz, which was also fine but required extra effort to ship specimens in the aftermath. All in all, it was relatively seamless given the breadth and scope of our group.

After carefully studying BioBlitz protocols, we improved on the recipe in six ways: (1) executed comprehensive whole-forest sampling from treetops to soil, not just surveys on the forest floor; (2) recruited experts who could classify all major groups of species, from water bears to fungi to slow lorises; (3) created strategic partnerships of Malaysian experts with international scientists; (4) diversified the team to include students, citizen scientists, government, university participants, and taxonomists; (5) invited women and minorities in significant proportions as participants; and (6) used social media to share the findings around the globe. The existing literature made it clear that no BioBlitz prior to this had tackled all those goals at once. It took a year to plan, including the organization of two scientific summits in Penang to ramp up local enthusiasm, then ten days to execute the survey and another two years to write up the results, with some new species still awaiting scientific classification even as I write, probably as backlog on some overworked scientist's desk. This is a perfect example of how taxonomy runs into a bottleneck and, sadly, explains why scientists bemoan that some forests will be destroyed long before their biodiversity is ever classified.

To prepare for the event, an intrepid team of local foresters was hired to survey the hillside, creating an essential list of tree identifications, important to document the location of an insect feeding or a bird nesting in a canopy. In Malaysia, government agencies play a huge role in any decision-making or land stewardship, so I also spent many hours in large rooms of (mostly male) officials discussing the nuances of a UNESCO proposal. The BioBlitz survey information would be provided as part of the nomination. In addition to the UNESCO dossier, the loan of certain specimens to foreign institutions for identification also needed careful paperwork to ensure legal transfers. It is critical to conform to local regulations involving the collection of specimens by foreign scientists, not just in Malaysia but also many other countries. All UNESCO nominations must be submitted by regional governments, so many agencies, as well as the university, were actively engaged in writing the submission at the time of our survey.

It did not hurt to have a prior friendship with the Sultana of Johor, one of Malaysia's major proponents of conservation. Three years before, I was her keynote speaker for the country's Conservation Day, hosted by the royal family and specifically by her Highness, whose official name I carefully memorized before attending: Raja Zarith Sofiah Binti Sultan Idris Shah, wife of Sultan Ibrahim Ismail of Johor. After the talk, Raja Zarith Sofiah publicly committed to raising the salaries of enforcement officers in the parks as an incentive to reduce poaching, making their jobs competitive against black-market payments for tigers. Malaysia's tiger population had fallen to less than 300 by 2014, yet poachers could still illegally get paid RM 50,000 for a tiger. She also urged her government to raise the penalties for poachers and encouraged nongovernmental organizations to promote conservation awareness. She continues to advocate tirelessly for biodiversity.

Malaysia is battling more than just tiger poaching with regard to its rain forest conservation. One of the greatest current threats is the palm oil industry, which netted $50 billion in 2017 and continues to expand. Indonesia and Malaysia account for over three-quarters of the world's palm oil production, although Brazil and Africa are entering the market. In Asia, enormous swathes of dipterocarps were cleared for this cash crop, sold for multiple uses in manufacturing as well as an edible vegetable oil. Most consumers don't even know they are purchasing palm oil—or what palm oil is, or what it costs Earth—which is used in thousands of everyday products from shampoo to fast food to soaps to plastics to baked goods. The introduction of biofuels was another death knell for Malaysian forests, because it caused a big spike of clear-cutting in anticipation of a commercial market. Once these tall primary tropical forests were cleared, vast plantations of oil palm (*Elaeis guineensis*), a species native to Africa, were planted as single crops in expansive monocultures. Fire is often used for clearing primary forests to plant oil palms, and over the past decade, some of the world's largest fires have occurred in Southeast Asia due to palm oil plantations.

Dipterocarps are the centerpiece of Malaysian forest ecosystems and were my original inspiration for pursuing tropical ecology. My master's thesis advisor in Scotland, Peter Ashton, was a world expert on dipterocarps, conducting much of his fieldwork in Malaysia. They are arguably the most economically important woody plant family in existence. While I shivered in the cold Scottish Highlands trying to measure birch bud expansion on my rickety scaffold, Peter was studying in the hot, sultry tropics, after which he returned to summers in Scotland (admittedly not always that warm!). Hearing the stories about those tall, charismatic trees growing in Malaysia's hot, humid forests, he got me hooked on the tropics. During graduate work in Australia, dipterocarps came back on my radar in an indirect fashion. My PhD graduate advisor, the renowned ecologist Joe Connell, and I mused for five years over the fact that some forests are dominated by a single species, including Australian stands of *Nothofagus moorei* and vast tracts of rural eucalypts, plus *Mora* stands in Trinidad, *Pentaclethra* in Costa Rica, and, of course, the dipterocarps in Southeast Asia. How can one species dominate so successfully in a world with over sixty thousand tree species? To most ecologists, the tropics are synonymous with high diversity, and temperate-zone biologists travel great distances to study low-latitude, highly diverse stands. But because Joe and I were carefully identifying every individual seedling and tree in different Australian plots for long-term research, we started noticing the existence of monodominant stands. I had just finished three years working on Antarctic beech, which occupied almost 95 percent of the cool temperate rain forest pockets. The only common factor Joe and I could find among these dominant species were their underground partnerships with certain types of mycorrhizae, which seemed to offer a competitive advantage. These fungi lived in association with specific species and provided an extra mechanism to absorb water and nutrients from the soil. Even for a canopy specialist, this belowground explanation was thrilling and transformed my perceptions about the growth dynamics of whole trees.

Joe and I coauthored a publication, hypothesizing that mycorrhizal associations enabled some species to dominate by outcompeting others. We cited examples in both tropical and temperate regions. After our paper was published in 1987, many students tested our theory in different sites (and found it correct!). Since then, new tools have been developed to study roots in greater detail, finding that trees "communicate" through their root hairs. In some temperate forests such as those of the Pacific Northwest, a mature or mother tree will share resources with her seedlings through underground connectivity. In a sense, those soils contain a complex highway of mycorrhizae and root hairs, transferring benefits between parents and their offspring. Tropical stands with low diversity have not been studied as extensively as sites with higher species diversity. This represents a bias, since most scientists who travel from temperate zones to conduct fieldwork in the tropics are seeking highly diverse forests, those different from the low-diversity ones back home. I call this a "temperate bias," and after musing about the single-species stands of Antarctic beech in Australia, I was thrilled to finally undertake fieldwork on dipterocarps, which exhibited a similar monodominance in Malaysia.

Then came October 13, 2017. All participants convened at dawn for the onset of the BioBlitz on Penang Hill. Ready, set, go! Everyone scrambled into the jungle to find critters and bring them back to our "base camp," which amounted to an outside setup of simple folding tables, microscopes, alcohol, lights, enormous spreads of food and snacks, and other equipment as needed to count every creature. One entomologist brought tiny inflatable swimming pools from which he created outdoor collection basins for trapping insects attracted to water. Another used special sieves to find micro-arthropods in soil samples. As vials were brought in from the canopy, taxonomists looked under microscopes and eagerly counted ants, scorpions, beetles, mites, nematodes, larvae, and other six- or eight-legged creatures. Eight climbers were engaged to sample the crowns and collect whatever the biodiversity experts requested. This created great excitement, especially when new records of arboreal ants and epiphytes were brought down from above. Visitors flocked to the Habitat eco-park and gazed in awe at the tables of scientists excitedly making discoveries. The methods of each team were quite different, depending on the taxon under investigation. Like the organisms they studied, many biologists became active at night, counting bats, spiders, or other nocturnal creatures. As some awoke early to watch birds, others were just going to bed after catching scorpions. The canopy team used single-rope techniques and pole pruners to collect leaf, flower, and fruit specimens from tall trees. They later pressed all collections for accession into the herbaria at USM, plus duplicates to loan elsewhere. With international biodiversity surveys, it is critical to lodge the specimens at a local institution, but sometimes duplicates can be loaned elsewhere. The ornithology team carried out point count surveys, meaning they stood in the same place at specific times of day, and recorded bird sightings and sounds. They also collected records and photographs for the iNaturalist website. Meanwhile, the ant team was literally crawling on the forest floor, juggling a pooter to suck up specimens, a machete to dig into rotten wood, and notebooks. The myrmecologists shared a few gadgets with other entomologists, including yellow pan traps (certain flying insects are attracted to both a water medium and the color yellow), pitfall traps, sweep nets, and light traps. Akin to the choreography of a ballet performance, the woods were alive with human bodies engaged in artistic motions but dedicated to biological collection.

A BioBlitz represents lots of perspiration, focus, and exclamation as scientists make discoveries in pursuit of their specific taxa. Overall, it was a hot and dirty process. We extracted and identified forty-three species of termites, requiring a microscopic follow-up in the laboratory to isolate each type into vials of alcohol with hundreds of tiny floating brown blobs, which may ultimately swirl into posterity in a museum drawer. We also cataloged eleven species of tardigrades representing the first collections of this phylum (Tardigrada) in Malaysia, of which two species were new to the world. To identify these creatures too small to see with a naked eye, Dr. Randy "Water Bear" Miller spent many hours poring over small bits of moist moss, leaf, and bark collections under a scanning electron microscope (SEM) looking for microscopic tardigrades, plus a year's follow-up to compare his photographs with existing collections around the world. The identification of 220 lianas and 300 Diptera (aka flies and their relatives) was absolutely incredible, as was the final count of 490 arachnids. Another prized sighting was the Sunda slow loris (*Nycticebus coucang*), a small nocturnal primate with enormous forward-facing eyes. It looked almost extraterrestrial when spotted in a headlamp late at night. Completely arboreal in behavior, the loris was christened as our BioBlitz mascot. Lorises are omnivores, feeding on vegetation, bird eggs, insects, and animal matter. As the world's only venomous primate, the loris secretes poison from glands on its elbows, which it licks before biting any enemies. Lorises are increasingly victims of the illegal pet trade, and because of their cryptic habits, accurate population counts are difficult.

By deploying the climbing team, who collected extensive material from the canopy, all scientists were able to survey species from the bottom to the top of the forest for each taxon. In addition, treetop leaf samples were collected from each species, creating a snapshot of defoliation for one point in time, not as accurate as monitoring leaves for many years as I have done elsewhere, but allowing an overview of insect-plant interactions. The JASON high school students measured insect defoliation by calculating the proportions of leaf surface eaten for samples of thirty leaves per species, height, and age. Herbivory ranged from 0 percent leaf area eaten (*Ficus* sp.) to 61.9 percent defoliation in one individual of *Cinnamomum porrectum*, with an average of 31 percent defoliation in *C. porrectum* throughout the hillside. After sampling hundreds of trees for a rapid assessment (aka snapshot herbivory) throughout my lifetime, Penang Hill was the first place in thirty years to find one species with absolutely no herbivory. Every leaf of the local fig (*Ficus* sp.) was uneaten, the aforementioned 0 percent! I can only imagine my Amazon shaman friend, Guillermo, excitedly asking about the special medicinal qualities of those incredibly toxic leaves. Unfortunately, we couldn't identify these tall figs to species, simply because the fruits were not present.

This BioBlitz advanced the natural history toolkit in several ways. First and foremost, our survey successfully deployed a whole-forest approach spanning from soils to the uppermost treetop, using eight professional climbers. Canopies were rigged with climbing ropes throughout the hillside, including our iconic and tallest species, *Shorea curtisii*, a locally common dipterocarp growing up to two hundred feet high. Second, the surveys included a large range of taxa: microbes, algae, tardigrades, arthropods, ferns, vines, trees, and vertebrates. During a ten-day span, we documented new field records: new species of ghost scorpion and tardigrades, new algae documented for the first time in this region, new records of ants in the treetops and rare ferns on rocks, and the discovery of ultrasound communication by the nocturnal

Sunda colugo (*Galeopterus variegatus*), detected at night when biologists were tracking bats. A new species of whip spider, *Phrynichus cockrelli*, was named after the visionary donors who built the walkways and funded the survey. Our total of 1,659 records were tallied at a symposium held on the last day of the expedition, although the numbers of some taxa will not be finalized until well into the 2020s.

Another distinction of our BioBlitz was its composition of participants, with local-plus-global collaborators, and the fact that 65 percent of our 117 participants were female, an absolute record for any rain forest expedition to my knowledge. (I think back to my being the sole female on the Cameroon hot-air balloon expedition and chuckle.) The survey also engaged global youth both in person and virtually, by hosting Malaysian and Hong Kong high school students to work side by side with scientists, and online livestreaming to K–12 classrooms. I had obtained a separate grant to rekindle the distance-learning JASON Project for its first tropical expedition since broadcasting from the Amazon walkway almost two decades before. Advances in technology had really changed the dynamics of virtual expeditions: the original JASON cost several million dollars to ship equipment and import a broadcasting team to film, edit, and relay a signal via satellite around the world; our reboot cost $50,000 for a laptop, three technical hosts, and a cadre of students partnered with a few teachers.

What was the secret ingredient of this phenomenally successful BioBlitz? We had great weather—one week after the expedition, it rained so much that half the hillside fell into the valley and closed the mountain for two months; those are the risks of fieldwork. We had wonderful compatibility in the field teams, so much so that a few groups are still collaborating on global projects. But the essential driver of our success was great government support as well as visionary funders, plus the pairing of local with international scientists. Those collaborations also garnered media attention, with praise from both government and corporate entities that enhanced the underlying goal of promoting conservation throughout diverse stakeholders. In ten frenzied days, we recorded over fifteen hundred never-before-officially-documented records on Penang Hill. What if every museum would undertake to send their collections staff to survey one unexplored ecosystem on the planet, and what if all museums made a coordinated strategy to each sample a different place? Scientists could probably advance our global knowledge of species diversity nearly tenfold. Field biologists often fall into a convention of returning to the same place and focusing on one major site throughout an entire career. As a result, some places become well studied while others remain underfunded and unexplored. Our BioBlitz was an example of an underexplored place yielding immediate discoveries; in ten short days, we unlocked new secrets of Malaysian tropical rain forests.

One of the unique attributes of Penang Hill is its location within a fifteen-minute drive of approximately two million residents, plus annual visitation by over three million tourists. Not many tropical ecosystems can claim such broad-reaching public access, making this site a great ambassador of conservation, citizen science, and education outreach. When finalized, the UNESCO proposal was entitled "Penang Hill Biosphere Reserve: Gem of the Island—Where Nature Is Conserved, and Culture Celebrated." A copy of the 188-page UNESCO dossier sits on my desk, and it also sits on a few government desks halfway around the world, pending consideration by an international committee. Penang has eight UNESCO cultural sites but no biological sites, so everyone was hopeful of success that will ensure this tropical forest is protected in perpetuity.

Another innovative feature of this site involved the green business model of the Habitat, a wildlife park with its two new ribbon bridges dedicated to research and education about tropical treetops. The Habitat's generous funders envisioned a canopy research station in the future. Fieldwork would be a solo effort no longer, as was the case when I climbed in Australia as a lone arbornaut. Our BioBlitz exemplified a new model of field research, with inclusivity from many diverse partners: the Habitat Foundation; Penang State Forestry Department; California Academy of Sciences; Universiti Sains Malaysia; TREE Foundation; Habitat Penang Hill; Department of Wildlife and National Parks, Malaysia; Penang Hill Corporation; University of

California, Berkeley; JASON Learning; Fulbright; PBS *Nature*; National University of Singapore; Tree Climbing Planet; World Wildlife Fund–Hong Kong; and Tree Projects.

The Malaysian canopy project taught me new ways to conduct the business of conservation. First, it saved time and energy to have a dedicated donor, rather than writing grants requiring multiple attempts over many years to obtain funds. Second, this project reinforced the notion that ecotourism is a critical conservation tool. The first walkway I helped design in Australia over three decades ago was a relatively simple structure to assist Earthwatch volunteers and a handful of guests at a small forest lodge in Queensland. These two Penang walkways have the potential to educate five million people per year. And third, building trust with the local community and including diverse stakeholders was key. In this case, the project included students, government officials, local university scientists, international experts, media, and Asian NGOs. This was not a formula I learned in graduate school but from the school of hard knocks, after many decades of watching deforestation and struggling to engage in more effective conservation. There is still a long way to go in this tree-saving business, but the formula for success is improving!

BioBlitzes are now a widely used, international tool for surveying species in different ecosystems. People who can't take an entire week away from work to travel to the Amazon jungle can instead take a day to survey a local redwood stand. Hosted in numerous parks, reserves, urban settings, and even backyards, such surveys often are limited to a specific taxon. The apps on cell phones to survey biodiversity are providing a platform whereby species discoveries are delivered seamlessly to the experts. With the app iNaturalist, people post images online and taxonomists can confirm their identification. All of this relies on volunteers, however, and the power of crowds to ensure accuracy and amplification of the data sets. A big benefit of any citizen science program is the engagement of a broader public in natural history. But similar to technical research, such actions all too often report on the presence or absence of endangered species but do not save them.

The scientific community, both volunteer- and professional-driven, still struggles to create actionable pathways to prevent the loss of species and habitat, rather than simply reporting. But by engaging the broader public in field biology—whether through BioBlitzes or online apps or simply the inspiration to get families outdoors to observe their local wildlife—the combination of science and technology can inspire stewardship of our planet's biodiversity.

⫸ Dark Red Meranti ⫷
(*Shorea curtisii*)

HOW MANY PEOPLE THROUGHOUT THE PLANET can recite the names of the world's most economically important trees, also one of our highest carbon-storing plant families? They are Dipterocarpaceae, a family with sixteen genera and approximately 695 species distributed predominantly in lowland tropical Southeast Asia. Their common name is dipterocarp, not part of the vocabulary of most people. The largest genus is *Shorea* (196 species), with other genera including *Hopea* (104 species), *Dipterocarpus* (70 species), and *Vatica* (65 species). The word *Dipterocarpus* originated from Greek: *di* means "two," *pteron* means "wing," and *karpos* means "fruit" (their fruits consisting of a hard, oily seed with one or two "wings"). The smooth, straight trunks reach nearly 300 feet tall, often without side branches for over 100 feet. The tallest known specimen is *Shorea faguetiana*, at 305 feet tall. Borneo is considered the epicenter for dipterocarps, representing 22 percent of all trees of some 270 species, including 155 endemics. Dipterocarps comprise the majority of biomass not only in Borneo, but also in Java, Sumatra, Malaysia, and wetter parts of the Philippines.

Dipterocarps are highly sought for their fine, straight timber used for plywood, furniture, flooring, boatbuilding, instruments, aromatic oils, and resins. One dominant canopy tree at our Penang Hill BioBlitz site was *Shorea curtisii*, commonly called dark red meranti. Our climbing team rigged one magnificent specimen tree for collecting as well as to train the local park guides for future arbornaut activities. *S. curtisii* is still selectively logged in Malaysia and Indonesia, though harvesting is restricted to large individuals with adequate numbers of conspecifics in the surrounding hectare. However, such legal requirements are not always accurately followed by loggers, nor is there much data on the inevitable decline of this species from harvesting. When left intact, dark red meranti provides critical habitat for many endangered wildlife: Bornean orangutans (*Pongo pygmaeus*), Sumatran rhinoceroses (*Dicerorhinus sumatrensis harrissoni*), Borneo pygmy elephants (*Elephas maximus borneensis*), proboscis monkeys (*Nasalis larvatus*), Borneo black-banded squirrels (*Callosciurus orestes*), leopard cats (*Prionailurus bengalensis*), and others.

S. curtisii is native to Malaysia, Indonesia, Borneo, Singapore, and Thailand, and one of the most common trees in these regions. As an emergent rising over two hundred feet in height, it has either gray or reddish-brown bark with coarse fissures. The leaves of *S. curtisii* are hairless, smooth, elliptical in shape, and approximately four inches long. Its small flowers are usually white or pale yellow, with five petals and fifteen stamens, although large enough for tiny insects called thrips to pollinate. Many dipterocarp flowers are highly scented to attract pollinators, either thrips or sometimes beetles. Fruits are winged, with three large and two short wings that turn red before ripening and eventually helicopter to the forest floor to germinate. Seedlings often germinate near adults because their winged seeds can't float far from the parent. As mast seeders, flowering occurs only once or twice every decade, and sometimes longer. Such infrequent fruiting is a temporal escape strategy because the rapid pulse of seed production makes it impossible for frugivores to build up enough population to consume the entire crop over such a rapid time frame. In Borneo, the El Niño/ Southern Oscillation (ENSO) triggers the flowering and fruiting events, so future climate disruptions could negatively impact regeneration. Like other dipterocarps, *Shorea* species have unique underground fungal relationships, whereby their root hairs are intertwined with ectomycorrhizae that confer a competitive ability to absorb more water and nutrients than trees without a fungal partner. In the tropics, where competition for both soil resources and canopy space is critical, having fungal associates called mycorrhizae is a real advantage.

In addition to the exciting underground world of dipterocarps, another wonderful *Shorea* story took place in the crowns, and involved the discovery of their pollinators. According to the story, two Sri Lankan graduate students erected ladders into a tall dipterocarp and established a treetop base camp to patiently figure out what pollinated these mast seeders. One night it finally happened, and as with all mast seeding, it happened all at once. Tiny thrips, about a millimeter (0.04 inch) long with tiny nondescript bodies and small fringed wings, descended by the millions, pollinating *Shorea* crowns in rapid succession. Both graduate students climbed down and soon after became engaged, a true romance in the annals of arbornauts! Since this momentous event, a few other pollinators such as bees, beetles, and moths have occasionally been observed, but thrips are critical for the reproduction of this economic and ecological keystone species. Not too much else is known about *Shorea* canopies due to their inaccessibility. Their leaves deter herbivores due to bitter-tasting tannins, and even the leaf-eating colugo will not take a bite. The fact that one handsome *Shorea curtisii* is now permanently rigged as an iconic climbing tree at the Habitat eco-park on Penang Hill will undoubtedly lead to more discoveries of its arboreal secrets.

10

BUILDING TRUST BETWEEN PRIESTS
AND ARBORNAUTS
⫸ ⫷
Saving the Forests of Ethiopia, One Church at a Time

S PARKS FLEW FROM THE JERRY-RIGGED GENERATOR chugging along as best it could, flashing Google Earth images of Ethiopia's remaining forest fragments, tiny green dots surrounded by huge expanses of brown subsistence agriculture. Dressed in khaki field clothes, I was an anomaly among an audience of over one hundred Ethiopian priests with flowing white robes and turbans. We had all come together with a common mission: to save the last native trees of northern Ethiopia. I had flown to Bahir Dar, Ethiopia, specifically to speak with the Orthodox Church leadership in the northern half of the country, where the only remaining native forests surrounded the church buildings; they were called church forests. These tiny fragments represented the last bastion for not only native trees, but also birds, insects, and mammals. Without these forest patches, the people's livelihoods without pollinators or fresh water would be tenuous, and their biodiversity would likely become extinct. The priests were curious, and I could only imagine their quiet discussions—why would a lone female, not a practicing Ethiopian Orthodox churchgoer, and a white person at that, make such an arduous trip to discuss their local forests? My job was to explain why these small fragments of trees were not only important to Ethiopia, but unique to the world. By now I was an experienced arbornaut, re-prioritizing my career to use my knowledge of trees to save unique species, forests, and their biodiversity. Ethiopia was a huge challenge given the extreme poverty throughout most of the country, which in turn has led to an overall lack of investment in science or conservation. But her trees and insects, given their endemic status and their essential function for the African ecosystems, are every bit as important to planetary health as those in California or Peru.

The Ethiopian Orthodox churches (technically called Ethiopian Orthodox Tewahedo Church, or EOTC) are surrounded by sacred groves; these are original (or primary) forests over a thousand years old. The churches dedicate their religious mission to protecting all of God's creatures, as well as the human spirit. These forests were roosting grounds for endemic birds like the hamerkops, several species of hornbills, and sunbirds—as well as a treasure trove of over 168 species of trees, which were documented in an Ethiopian student's PhD thesis dated 2007. As a conservation scientist, my mission is the same, but simply uses different vocabulary: to protect biological diversity.

My old laptop sputtered with the frequent electric surges, but it held up. When the priests saw aerial images of their own forests, surrounded by brown dirt and dry croplands, they gasped. They had no access to computers, Google Earth images, or even a biology book to help them learn about island biogeography, which is an ecological concept applicable to forest fragments. Depicting these images in my talk helped me convey the urgent need to forge a partnership between religion and science, to conserve these green treasures. Ethiopia was one of the most extreme examples of deforestation I'd ever seen.

Imagine approximately thirty-five thousand small stands of several hundred to many thousands of trees amid a barren landscape of subsistence agriculture. These green patches ranged from five to six hundred acres in size, and the number itself was highly debated because no one knows how to define a fragment—is it ten trees? Or ten acres? Some estimates placed the number of northern Ethiopia's forest fragments at fifty-five thousand and others as low as twenty-one thousand. But we do know that less than 3 percent of the original 42 percent forest cover remains. This translates to under a million hectares (just over two million acres) of green dots surrounded by twenty-five million hectares (about fifty-five million acres) of subsistence agriculture. The tree stands are called church forests, and for thousands of years, local communities believed their house of worship should be surrounded by forests to provide refuge for all creatures. However, if your children are hungry or drought lowers your crop yield, it is understandably tempting to utilize this resource for survival. The locals occasionally prune trees around the perimeter for firewood, allow their cattle to graze on seedlings and understory foliage, overplant coffee crops in the understory, ring-bark older trunks to fell for additional timber, and sometimes use slingshots to secretly hunt mammals and birds when the family is desperately hungry. The priests lamented the degradation of their sacred forests, but without major economic or government influence, their only recourse was to pray for a solution; they had no influential budget, nor did they have the savvy to navigate politics. Due to their monastic lifestyle, many were not aware that the surrounding valleys had been clear-cut until they saw my aerial images in the presentation. These photos stirred the religious leaders into action, launching our unique partnership at several levels: science plus religion, one female working with hundreds of male priests, an American Christian among thousands of Ethiopian Orthodox villagers, and the integration of highly developed aerial technology with one of the world's oldest spiritual philosophies.

How did I discover this urgent situation in Ethiopia? In my seventeenth year as treasurer of the Association for Tropical Biology and Conservation (ATBC), our international meeting was held in Morelia, Mexico. I award a small check to the prize-winning student paper at every conference, and that year it went to Dr. Alemayehu Wassie Eshete, a recent PhD graduate whose thesis included a survey of native trees throughout northern Ethiopia in conjunction with the International Union for the Conservation of Nature (IUCN) endangered species list. This IUCN Red List earmarked species whose declining populations required urgent action. After handing him a check, I offered congratulations and politely asked, "What's next?" He nearly burst into tears, explaining he was the only person working on this urgent issue. How could I walk away saying "Good luck" and not offer mentoring or collaboration? After all, I had almost thirty years of canopy research under my belt and had devoted the last two decades to forest conservation in three other continents. Why not apply those years of experience to Ethiopia? During past fieldwork in the Amazon and India, despite my expanding CV, I had not directly prioritized saving trees, which caused me to rethink my actions in mid-career. I had taken personal action, twice leaving the comfort zone of academic tenure to join a museum in an effort to reach diverse audiences through a public platform. I also wrote some popular natural history books, pursued a lively calendar of talks ranging from middle schools to college commencements, and pioneered virtual expeditions for middle school students. These experiences served me well as I increasingly focused on forest conservation, not just academic output.

My new Ethiopian colleague, Alemayehu, had a nonconventional background, in the worlds of both religion and science. As a child, he studied to become an Orthodox priest, serving as a child disciple to a senior priest as well as spending long periods of solitude in the forest. After a decade of religious training, he observed the disappearance of the ecosystem immediately surrounding the church, which he had come to love. His parents convinced him that an advanced education would give him expertise in both religion and ecology. He recognized that his conservation goal was not achievable from the altar but needed a combination of religion and science, so Alemayehu pursued and obtained a scholarship to Wageningen University in the Netherlands to study forest ecology. There, he dedicated himself to this research effort, alone in a foreign country far away from his wife and children. He brought the same determination to our collective efforts for conservation. Alemayehu and I reinforced each other with an inspiration: What about tackling one of the world's most urgent forest challenges, saving the last forest fragments in Ethiopia called church forests?

Australia has a much higher level of scientific investment compared to many African countries. Could efforts to engage environmental stewards succeed in countries with fewer resources? And could conservation succeed with religious partners instead of academic or regulatory agencies? In Ethiopia, Alemayehu had little access to technology, fundraising, or sustainable solutions. That made it tough for him to convince local communities how important native forests are to human health. But he had earned the trust of local priests. For my part, I had access to technology, fundraising, and the latest scientific findings about ecology. We spent many hours of coffee-drinking in Mexico discussing the logistics of how the two of us could avert Ethiopia's rapid landscape degradation. Drinking a final cup of coffee, we pledged to work together to reverse the losses of native forest in Ethiopia using the very disparate assets of religion and science. Forests are spiritually critical to several billion people worldwide; this may prove a key driver in the future of global conservation. It is relatively easy to translate timber, water, or pollinators into dollar values to calculate the benefits of trees to humankind, but tougher to estimate the worth of spirituality. Maybe the number of prayers should be considered an important metric, ultimately giving more standing to the notion of retaining forests instead of harvesting them?

For his PhD thesis, Alemayehu measured the growth and distribution of 168 tree species in 28 local church forests, earmarking those classified by international conventions as having endangered status. He observed their declining health due to human activities, especially the excessive grazing by livestock that invaded to eat juicy seedlings and understory foliage. He had learned how to calculate stand density and seedling decline in small plots, but he had no direct solutions for the priests to reverse the situation. Most families struggled to put food on the table and were not focused on the insidious degradation of native plants. And none of the priests had the scientific training to measure perimeter shrinkage or overgrazing of their church forests; they had only their observations that the integrity of their trees was increasingly compromised. After Alemayehu obtained his graduate degree in 2007, the priests started criticizing him about his research, because they realized his simply listing species and writing reports were not actions that directly saved trees in their sacred sanctuaries. They were right. He had identified the forest decline but had no positive actions to avert the situation. Even more discouraging in his eyes was that local women, ideally suited as community environmental stewards because of their daily chores, were not part of any solutions. Girls often left school after fifth grade and had their first child at age thirteen, but they had no voice as local stewards of their environment, despite their interactions with nature to fetch water, collect firewood, and grow gardens. Alemayehu was despondent over the fate of his country's shrinking forest fragments, frustrated by the societal inclination to exclude women's knowledge from environmental solutions, and dismayed to find no government support and not even a local NGO partner.

As a first step to acting on my pledge in Mexico, I bought Alemayehu a plane ticket so he could participate as a visiting researcher in the environmental studies program at New College, Florida. Our family was thrilled to house him during his stay. One night, while surfing the web on our family computer at 2:00 a.m., he discovered Google Earth. I awoke to hear the printer chugging away and Alemayehu mesmerized by aerial images to share with the priests. They depicted jagged perimeters where farmers' plows had made intrusions into the church property, gaps in the canopy where the red stinkwood trees had been illicitly cut, and excessive walking trails crisscrossing the landscape. Google Earth confirmed Alemayehu's ground-truthing observations: northern Ethiopia's remaining native stands were in serious jeopardy. Because few priests ever left their small green enclaves, they did not recognize the severity of the degradation throughout adjacent valleys. And similarly, their local parishioners did not travel far afield to understand the broader context of Ethiopia's deforestation.

After Alemayehu's visit to Florida in 2008, I spent the rest of the year learning about African plant conservation from the existing (limited) literature. I read and reread Alemayehu's thesis, plus a small handful of other publications, many of which focused on forest restoration using non-native species. The introduction of Australian eucalypts by the Ethiopian government dominated most reports. Almost like snake oil solutions, gum trees were touted as a possible canopy restoration for Africa. But they represented an insidious threat to the native canopy for three important reasons: (1) although non-native gums grew faster outside of Australia in the absence of their natural pests, their timber was not only inferior to the local species, but also a fire hazard due to their fire-prone chemistry; (2) their canopies excluded native biodiversity due to volatile eucalypt oils; and (3) they required about four times more water than native Ethiopian trees, sucking down the water table to dangerously low levels. We needed two quick actions: exclude the livestock from the church forests to save the seedlings and understory foliage, and eliminate the invasion of non-native eucalypts. We determined that an effective solution was to build conservation walls to protect the groves. Initially, I thought back to the success of barbed wire used for our tree planting programs on the Australian rural landscape. Maybe simple barbed-wire fences could protect the church forests? I tried to convince a large agricultural corporation to donate enormous spools of barbed wire for temporary fencing. No luck!

The following year, I booked a ticket to Ethiopia and headed off with a small carry-on bag to see the sacred trees firsthand and further strategize with Alemayehu about an action plan. It took twenty hours of flying—Tampa to New York, New York to Frankfurt, Frankfurt to Addis Ababa, and Addis to Bahir Dar, the gateway town for northern Ethiopia. On the last leg, I pressed my face against the airplane window, straining to spot any isolated green dots similar to the Google Earth images. They were few and far between. The landscape was much more barren than I ever imagined. Gripping the armrests, I wondered how to remain optimistic despite such a daunting challenge. Alemayehu was convinced most of the forest decline had occurred over the past fifty years, but because there were no written records, we needed to ask the priests directly about the recent history of the invading livestock and eucalypts. For over a thousand years, these Orthodox enclaves were icons of faith across the Ethiopian highlands. But during the twentieth century, without irrigation or metal tools, crop yields remained extremely low, making it necessary to clear more, and then more, land to feed families. Alemayehu learned from his thesis research that if Ethiopians lose those last remaining native fragments, they will also lose their freshwater springs, biodiversity, soil conservation, shade, honey, home for many pollinators, carbon storage, sources of dyes for religious murals and fabrics, seed banks, future timber, and, perhaps most important, their spiritual sanctuaries.

Many of Ethiopia's native trees are endemic, meaning they are found nowhere else in the world. The incredible African cherry, or red stinkwood (*Prunus africana*), is not only noble in stature as the tallest species of the family Rosaceae, but also serves as a medicinal plant throughout Africa. The IUCN Red List of threatened or endangered species includes many Ethiopian species: baobab (*Adansonia digitata*), coral tree (*Erythrina brucei*), and broom cluster fig (*Ficus sur*), to name a few. Equally important, the red stinkwood canopies are home to many native birds, animals, and insects representing the country's biological heritage, including economic insects that pollinate crops, unique birds that are icons for tourism, and native mammals. These last patches of forest represent a Noah's ark for Ethiopia, supporting her genetic library of species essential for future ecological restoration. Given the accelerated clearing of Ethiopia, confirmed from observations by locals, there was not much time remaining before these genetic libraries would disappear forever. Many remaining fragments were already degraded significantly, requiring urgent efforts to reverse the impacts of cattle grazing, firewood gathering, excessive plowing around the edges, illegal hunting, and invasive eucalypts replacing the native canopy. As the plane touched down in Bahir Dar, I really had no idea what to expect. Could Alemayehu and I achieve our lofty goal to save these precious church forests?

As our first step to creating a solution, we needed to communicate with the religious leaders and develop a sense of trust between scientists and clergy. So we hosted a workshop where we could make a visual presentation of our Google Earth images and lead a serious discussion about forest degradation. I also made bound pamphlets of the images to hand out to each priest. Typical of field biologists, I selected a ridiculously cheap hotel as my base camp in Bahir Dar. There was no cell phone app to advise me, only a Lonely Planet guidebook that listed the Ghion Hotel for less than ten dollars per day. Water (non-potable) barely dripped from my rusty tap, so I set a cup in the tub and when it was full, managed to "shower" using a wet cloth. Electricity was limited to a single bulb dangling from the ceiling, and the linens did not appear to have been washed since they were purchased, probably decades ago. Guards roamed the paths to ensure the homeless did not wander into our open-air rooms. On the plus side, the hotel's coffee was wonderful. Coffee originated in Ethiopia. There's a delightful legend of early goatherders who observed their goats jumping and energized after feeding on the wild berries of an understory shrub. The next time the herders got tired, they chewed on the berries to revive their own energy levels, and coffee was born. On the first morning, I awoke to raucous squawks of silvery-cheeked hornbills (*Bycanistes brevis*), with their oversized beaks, joyfully frolicking overhead in figs, whose massive canopies sprawled over the entire hostel grounds. The bird calls competed with Muslim priests who chanted loudly at dawn every day, blasting prayers via loudspeakers throughout the streets.

Every morning during this first 2009 visit, I relished a hornbill wake-up call and drinking Ethiopian coffee while wearing mittens in the frosty hotel garden. The first day, after breakfast in that chilly January morning air, Alemayehu took me to visit a church forest called Zhara, about fifteen miles outside Bahir Dar. After driving past brown hillsides of subsistence agriculture, we arrived at a green oasis totaling eighteen acres along the roadside. The contrast was striking. Inside, singing birds and buzzing insects surrounded us with their symphonies, whereas the bare fields outside were silent. Rich, botanical odors permeated the shaded sanctuary with the beautiful aromas of flowers, decaying wood, foliage, and bark. A few hamerkops (*Scopus umbretta*)—an ungainly prehistoric-looking bird—flopped around in the treetops, and dozens of smaller birds sang their hearts out in the lush greenery. Under a leafy canopy, we walked past a freshwater spring, where several people were bathing in the healing elixir. Tree stumps indicated where the locals had butchered precious crowns to let the sunlight penetrate on patches of coffee in the understory—all sustainable activities but so much expanded in scope that they now jeopardized forest health.

I was humbled to meet Yeneta Tibebu, the priest of Zhara and one of the most influential religious figures in the region. He was very shy (just like me) and spoke very softly, but permitted me to bow down and kiss his cross, a trusted exchange of friendship and respect. Alemayehu invited him to our workshop, and he quietly agreed to attend. While traipsing around the edges of Zhara where dry grasses merged with trees, I unintentionally attracted an unwelcome onslaught of biodiversity. Two days later, I had twenty-six chigger bites and counting. The bites had come out by the first evening and throughout the next day, as those voracious critters romped throughout all my warmest interstices and chomped away, leaving their itchy toxins in the wounds. These biting mites especially loved underwear lines, under the elastic! Oh, the joys of fieldwork in dry grass. It took a week for the fierce itching to die down, and on the next visit, I stayed within the bounds of the shaded vegetation, where chiggers don't lurk.

Our workshop was scheduled two days later, in a rural town called Debre Tabor, which was also the regional headquarters of the Orthodox Church. We left in a rental jeep in the predawn hours, first picking up the priest at Zhara and then stopping at another green oasis, eight acres called Wonchet, to offer a ride to their chief priest, Aba Tewachew, who proceeded to sit in the back seat and throw up. Alemayehu quietly explained he had never ridden in a motor vehicle before. Even though I had traveled widely, I found the landscape of eastern Africa unlike any other. Isolated figs and stinkwood dotted an otherwise brown landscape, where teff would soon be planted as a seasonal grain crop. No birds or wildlife inhabited the expansive agricultural lands interspersed with small mud huts covered by thatched roofs. People walked along an extremely hot and dusty dirt track heading to market, sometimes in groups with animated conversations. Women carried grain or produce in cloth sacks on their heads, to trade for other goods. Occasionally, a family hauled an entire cowhide like a giant, stiff board; one man hauled a beehive affixed to a branch. The concept of walking miles to market, carrying your wares and your shopping on your head, is inconceivable in industrialized countries. It's difficult to envision how the world will achieve equity for women in the near future given such extreme economic divides.

Eventually, the main highway turned into a dirt road, one of the bumpiest tracks I've ever experienced, with enormous swirls of dust blotting visibility for the driver as well as for the people walking along the roadside to market. I choked from inhaling so much Ethiopian soil, and all my clothes were coated in a thick layer of brown when we finally arrived at Debre Tabor. Our driver approached a bright-pink hotel on the side of the dirt road, with a cluster of mud-encrusted trucks out front having their tires changed. This was our new base camp. Although the gaudy coat of bubble gum–colored paint made the facade appear new, that was not the case for the interior. I was escorted to the third floor to a "room with a view" overlooking the dust, tire-changing, and stalls where kids bought Coke and chewing gum. The hotel water supply was temporarily off, which turned out to be permanent, and the electricity went off about an hour after we arrived. I was grateful to have a headlamp to navigate up three flights of stairs and opted out of eating local fare in the dark, collapsing into bed wearing my dirty clothes.

Alemayehu had rented a hall, plus paid a driver with his rugged jeep to transport us and our equipment from Bahir Dar to Debre Tabor. He wanted to make a good impression on the priests because none of them had ever attended a church forest ecology lecture. The first hurdle had been to convince the religious leaders to attend, and I'd worried when Alemayehu had advised me that each one needed a per diem as incentive. We had no grant funding, and I had already purchased an expensive airline ticket and paid for 150 color booklets, but I knew he was right—the guaranteed participation of these religious leaders was essential. Alemayehu had further explained that I needed to pay them for three days: one day to travel to the workshop, one to attend, and another to return, plus meals, hotel, and bus fares. I'd fretted over how to swing these extra costs, thinking of the average per diem costs for conference attendees in North America. But Alemayehu's calculations wound up translating to $10 per priest, illustrating the vast difference in cost of living between rural Ethiopia and other countries.

I was relieved to write a check for $1,500 to bring 150 priests to our workshop.

The priests did not let us down. We had standing room only at our first workshop, and through some celestial miracle, the generator worked. Alemayehu translated my presentation into their native language. Though there are a handful of women's monasteries in the region, most clerical leadership in Ethiopia is male, so I was the only woman at the table. They gasped at the aerial images of their landscapes and were shocked to see so much of the surrounding region cleared and replaced by cultivated fields of teff, millet, and corn. The priests engaged in lively discussion, criticizing their government's program to plant eucalypts and fully recognizing how much water those non-natives sucked out of the ground, as well as their toxicity to local biodiversity. Increasingly animated, these 150 men, ranging in age from twenty-five to eighty-five, began discussing solutions ranging from conservation walls to removing eucalypt seedlings to restoring native species. That day, the priests pledged to work together to save their remaining forests, recognizing that a unified voice would carry more weight with the local communities.

Science and spiritual faith are often cast as adversaries, but this is not the case in Ethiopia. With science as a driver, the Orthodox priests are now engaged in aggressive conservation practices to restore their church forests. Over two billion people throughout the world respect and protect sacred groves as part of their spiritual heritage, and over 1.3 billion people live on degraded agricultural lands, so perhaps Ethiopia's solutions can apply to other countries. Working through the powerful force of religion, perhaps conservation can gain leverage where governments and corporations cannot. Soon after the workshop, the bishop of the northern Ethiopian diocese awarded me the first official memorandum of understanding (MOU) for an outsider, allowing me permanent access to the church forests. This privilege was accompanied by the bishop's personal plea to me to never give up on our shared mission, recognizing that we were more powerful as a team. After thirty years as an arbornaut, my fieldwork on Ethiopian church forests reminded me of what I had learned in the Australian outback many years ago: that building trust was the most valuable asset to achieve bottom-up or community-based conservation.

The Ethiopian Orthodox Tewahedo religion prided itself on being older than its Egyptian Coptic counterparts. Rural communities engaged in daily prayer sessions, frequent sacred events (birth, baptism, marriage, death), and daily events that benefited from a clerical blessing. Situated in the center of the church property, the Ethiopian Orthodox building is always round and contains three circular internal sections, symbolizing the Trinity. Most forests surround a spring as a source of both drinking and healing waters. As a scientist, it was not normal for me to spend extensive periods of time in prayer, but as a partner in Ethiopian conservation, I quickly adapted. The best way to work seamlessly with locals is to embrace their cultural activities. I partook in ceremonies such as eating freshly roasted lamb while sitting on the forest floor, drinking homemade hooch from six-gallon drums, eating stews whose contents were unidentifiable, and praying between every bite. These rural people exuded a boundless joy and reverence for life. In fact, I was almost jealous watching fathers plow the fields with their small boys in tow and the women fetching water with their daughters.

Local people utilized medicinal plants from the church forest trees for most of their ailments. The iconic African cherry (*Prunus africana*) was greatly overharvested because it allegedly cured prostate cancer, despite having malodorous fruits that had earned it its other common name, red stinkwood. Other medicinal trees included figs (*Ficus sur* and *F. vasta*), whose milky juice kills intestinal parasites (similar uses exist for *Ficus* in the Amazon); hagenia or kosso tree (*Hagenia abyssinica*), used for tapeworm infestation; corkwood (*Commiphora schimperi*), which was the original source of myrrh; and baobab (*Adansonia digitata*), with numerous medicinal purposes. The sacred African juniper (*Juniperus procera*) was sustainably harvested for construction, but only used for churches given its history as the timber selected for Jesus's crucifixion cross. Most priests carved a cross of juniper for their worship ceremonies. Unlike the Amazon region where ethnobotany has been fairly well documented by visiting botanists, few publications exist about the medicinal uses of Ethiopian plants. Instead, people rely on oral history, and the science is still catching up.

Most native trees grew more slowly than the introduced eucalypts because Ethiopian species lived amid their natural enemies that keep their growth in check. This factor alone led to a temptation to plant non-natives. But locals also observed that eucalypts depleted the water table by taking in approximately four times more water than native trees, as well as excluded native species from their canopies due to their leaf and bark toxicity. So even without formal ecology training, the priests called them "evil trees." Although slower growing, native trees were evolutionarily adapted to the localized poor soils and frequency of drought, and their canopies represented the major habitat for endemic animals. Vervet monkeys (*Chlorocebus pygerythus*) cavorted in the mid-canopy along with rock hyrax (*Procavia capensis*), while understory residents such as bush duiker (*Sylvicapra grimmia*), hyena (*Hyaena hyaena*), and the spotted cat called the Abyssinian genet (*Genetta abyssinica*) relied on the native understory for habitat. My life list of new Ethiopian birds exceeded one hundred, including the rare yellow-fronted parrot (*Poicephalus flavifrons*), which depends upon the native trees in the Lake Tana region. If these last groves disappear, native biodiversity may become extinct.

I returned a year after the first workshop to host a second. Yeneta Tibebu, the forward-thinking priest of Zhara, was excited to share a surprise with me. Of his church's own volition, they had hauled rocks out of local fields and created gorgeous drystone walls around the church forest, a 1.6-kilometer (1 mile) perimeter safeguarding forty-six species of native trees and untold other biodiversity. This local conservation wall was a clever solution because it excluded cattle and goats, defined the forest boundaries for farmers, minimized firewood collection around the edges, saved the seed bank and biodiversity, and created a great source of pride for the local people. They referred to their stone walls as "clothing for the church," and they adopted a fervent belief that a religious enclave without a wall was incomplete. The stonework was about four feet high, two feet wide, and needed no maintenance, and in some cases, that wall was laid several yards out from existing forest boundaries to compensate for shrinkage of the perimeter over the last few decades. Several other church forests soon followed the Zhara model within a year after our inaugural workshop: Debresena built 1.9 kilometers (1.2 miles) of walls protecting forty tree species, and Mosha constructed 2.8 kilometers (1.75 miles) of wall surrounding fifty-three tree species.

Alemayehu explained to me that we should pledge a monetary donation for each completed wall, to help the priests purchase gates (four in total, one on each side) and to serve as an incentive. Since the priests have no cash economy, this donation could also assist with roof repairs or protection of valuable religious murals, all too often exposed to harsh weather. I now quietly raise funds as honoraria for Ethiopian conservation walls, and Alemayehu is the local broker who allocates the donations. Our efforts are greatly appreciated by the priests, leading to an enormous level of trust among us. Many international donors discover this project on my website and are motivated to donate after seeing these isolated fragments of green amid a brown expanse of subsistence agriculture. We budgeted the total cost for stone walls around forty forests at approximately $500,000, plus annual workshops, making this one of the cheapest countrywide biodiversity conservation programs in the world. We have raised over half the funds required to achieve our goal, and to date the majority of donors are elementary schoolchildren half a world away, who either sponsor a penny drive or a bake sale. Whenever I give a virtual "meet a scientist" talk to schools in North America, the kids usually follow up by donating a class gift to save trees in Ethiopia. One of the most thoughtful donations came from a low-income sixth-grade class in the Bronx, who gave a week's worth of their lunch money. A family in Germany donated a memorial tribute to their deceased father. A young Canadian girl set up a lemonade stand to fundraise for walls. Thanks to virtual tools and social media, kids can advance conservation on the other side of the planet! Scientific colleagues often debate whether connectivity corridors can be envisioned after the walls are completed, because they doubt the local people will ever expand or alter the original wall locations. Good news regarding connectivity between church forests! Two priests have

already expanded their wall boundaries to facilitate additional restoration, because local people are more than willing to provide labor to
shift the walls from their existing locations to expand the church forests. In the world of conservation, connecting different fragments of
habitat is greatly beneficial because it creates a larger refuge. Ethiopia
still lacks modern irrigation, tractors, and other advances to raise agricultural yield. For now, the local people need expanses of cropland to
grow enough food.

Ethiopia is a complex backdrop for conservation. As one of Africa's
most populated countries, it now exceeds 100 million residents, and
has abundant, unique natural resources—the headwaters of the Nile,
our human ancestors (home of the hominoid Lucy), strong religious
values, a wealth of minerals, and a unique endemic biodiversity found
nowhere else in the world. My local partner, Alemayehu, frequently
touted this list of Ethiopia's natural treasures, but he also felt frustration that his homeland never leveraged her resources as strategically
as other African countries. But Ethiopia is rapidly catching up. The
government recognizes that climate change is taking a toll on its landscapes with increasing levels of drought, locust infestations, and other
hardships. In one forward-thinking action, the president declared a
tree-planting day in 2019 when approximately 352 million seedlings
were set out onto the landscape, after germination in government-
funded nurseries. This outpaced any other country in the world to date,
and despite the high mortality those seedlings will undoubtedly face
lacking in water and horticultural maintenance, it shows extraordinary
conservation vision by a national leader. Ethiopia increasingly exhibits
actions that indicate a green future. I will continue to pray with the
priests, host workshops, and merit the trust of the clergy as we partner
between science and religion.

Once we had a firm action plan in motion to conserve the church
forests, the priests were more open to Alemayehu and me conducting
field research. They know that we will not simply collect data and document species, but that we are committed to solutions to save their trees
and "to protect all of God's creatures." As next steps, Alemayehu and
I embarked on a series of National Geographic–funded expeditions to

study the biodiversity of the church forest canopies. First, we wrote a grant to survey insects, funding a dozen international experts on different arthropod taxa. Our grant was successful! I sent out a request for adventurous entomologists to join our expedition and was overwhelmed with positive responses, including experts on ants, beetles, mites, flies, nematodes, and other six- and eight-legged critters. We rigged ropes in tree canopies, found new species, logged new distribution records, and engaged an enormous following of local schoolchildren who wanted to learn about their local bugs.

To survey insects, we created transects through two different church forests and used multiple methods: beating trays, light traps, sweep nets, bait traps, malaise traps, pitfall traps, and ordinary observations. The church forests are essentially small green fragments that represent incredible natural experiments of island biogeography. This allows us to compare species diversity and abundance between sites of different sizes, tree composition and ages, and elevations. The fieldwork was incredibly time-consuming, however, as compared to working in other countries. We needed to hire drivers, load up four-wheel-drive vehicles, purchase food, meet with each priest to approve our sampling, and then rig trees with canopy insect-sampling devices. All this preparation took the better part of several days before we caught one bug, and our efforts were not without hurdles. First, we set up our ropes for both climbing and setting insect traps in the canopy, which extended up to one hundred feet tall, but the next morning, all the ropes were carefully extricated from the branches and gone. We were able to buy more rope in the village, but after that mishap, Alemayehu hired guards for any gear that remained overnight in the trees, because local people joyfully removed the string and plastic vials, taking them home for more practical uses, probably in the garden or to tie market goods. After that, we reworked our budget to hire guards to watch our traps twenty-four hours a day—it wasn't expensive, but finding available men (women were not allowed to sit in the groves all night) took another few days.

All our fieldwork needed to respect ongoing church activities, so we carefully inquired about the times of funerals, church services, and other religious activities. We remained extremely quiet in the woods when a religious procession wended through the trees, and avoided taking photographs if the local people were present. Throughout our expedition, we hosted a large audience of onlookers so decided to utilize that opportunity to educate local students about their biodiversity. By the end of some sampling days, dozens of kids (mostly boys) trailed our scientists, so it was apparent they really wanted to learn more about their local natural history. According to Alemayehu, in the absence of computers, field guides, or textbooks, the schools offered little environmental education, and the teachers did not conduct field trips. I thought long and hard about this global inequity of school resources and the concurrent challenge to educate the next generation of Ethiopian priests (and biologists). So I piloted two ideas. The first was to make T-shirts depicting important native pollinators, labeled in the local language, Amharic, to give children an insect field guide on their backs. This had the added benefit of providing school clothing; many boys did not own a T-shirt, wearing blankets to school as ponchos instead. The T-shirts were also practical for lasting fieldwork; paper checklists would simply get wet or end up in mom's fireplace. The T-shirts were a huge hit, and even the priests wanted to wear them under their flowing white robes. Our volume is limited due to the expense of transporting them to Africa, but I am hopeful we'll soon find a local manufacturer to ramp up our environmental education fashions. As a second innovation for educating the local schoolchildren, we created a kids' book to illustrate why trees are such a valuable local resource. With an Ethiopian colleague, I coauthored *Beza—Who Saved the Forests of Ethiopia, One Church at a Time*, the story of a girl who helps conserve her local trees. Whenever someone buys an English copy of *Beza* on Amazon.com, my small foundation (www.treefoundation.org) donates an Amharic copy to a child or school in rural Ethiopia. It has been a humbling experience to give a girl (or boy) her first book, printed in her own tongue.

To set up insect traps, hire guards, work around the religious activities, and then collect/dry/pin/label the results took one full week for each site, so our two-week expedition did not allow sampling in ten church forests as we had hoped. But we did manage to collect great data for two sites, and since these were the first field surveys of arthropods in church forests, we were excited about the findings. In the end, only four collection methods (light traps, malaise traps, sweep netting, and bait traps) provided consistent sampling data to compare. We amassed 8,200 arthropods, including 253 morphospecies of Coleoptera (beetles) representing 37 families, and at least 45 species of oribatid mites. A particularly ugly, brown, hairy new species of mite was later named *Pilobatella lowmanae* in honor of the expedition leader by a Russian mite expert (although my children tease me, wondering if this scruffy-looking critter is really a tribute).

Our second National Geographic–funded project involved surveying the reptiles and amphibians. This time, Alemayehu and I proposed to conduct fieldwork in both the church forest vegetation and the conservation walls, wondering if the stonework might serve as new habitat for lizards and snakes. Instead of bringing a large team for one expedition, we enlisted a single graduate student who made repeat visits to both walled and unwalled forests. Almost no published information about herpetological biodiversity existed in this region, so our findings added new records to the halls of science. We also confirmed that the stone walls represent a safe refuge for many reptiles and amphibians. In fact, the majority of species surveyed were located in the interstices of the stones.

For over ten years, Alemayehu and I have worked from a conservation model that required one major ingredient: trust. It blossomed because we also have a true partnership: each of us brings something essential to the collaboration. I could never have gained the friendship of the Ethiopian Orthodox Tewahedo Church without a local colleague respected by religious leadership. And Alemayehu could not easily have taken such bold actions to save his forests without an outside partner who engaged fundraising and international attention. He is the ambassador to the locals; I am a liaison to the outside world. We continue to fund new conservation walls, host annual workshops for new priests throughout different regions of northern Ethiopia, distribute books and T-shirts, and share the importance of Ethiopian church forest conservation as a global priority.

In 2018, I visited Gibstawait, a newly walled forest with the largest recorded diversity of native birds. Upon my arrival, the priest and his disciples were overjoyed and immediately sacrificed a lamb (an infrequent event in this drought-stricken landscape). Within two hours, we were immersed in a true lovefest celebrating the successful protection of their sacred grove. We feasted on boiled lamb under the canopy, raised up prayers, and sipped their special alcohol (with which they refill glasses beyond overflowing). The religious community honors Alemayehu for his conservation solutions, a veritable rock star for the trees. My own sense of the meaning of global stewardship is deepened by working alongside the humble, albeit strong, people of this region. Our story shifts a paradigm for changing conservation, away from top-down government decision-making to dedicated bottom-up actions of local communities. But one big unknown remains: Is it too late? Are the forest fragments in Ethiopia too small to endure and will the diverse composition of native species be sustained? Do the isolated populations of trees and mammals have enough genetic variation to survive the onset of climate change? What is the fate of the medicinal red stinkwood given that only one or two remain in each stand?

I returned to Ethiopia as a Fulbright scholar to continue my focus on forest conservation, tree canopies, and women in science. Assigned to Jimma University, I worked on several initiatives to bring new ideas to the country's largest institute of higher learning. First, I taught a climbing class for the biology department. Sadly, only male students participated. While rigging the ropes, baboons competed with us for canopy space, throwing twigs at their primate cousins struggling to develop their skills in arboreal pursuits. Throughout the climbing activities, women carrying large cloth sacks picked coffee in the understory. The gender imbalances in Ethiopia remain challenging, so I started a university networking group for female STEM faculty. Only six women showed up, and they were all technicians. They explained that women can't be serious contenders for an academic job if a man has applied. My third assignment was to advise on a future botanical garden for Jimma University as the first plant museum for the southern region of the country. We selected a site, discussed ideas for public outreach, and crossed our fingers that such public sanctuaries will be part of Ethiopia's future.

What next? At the time of this book's writing, Alemayehu and I passed the halfway point to fund conservation walls around the forty highest-biodiversity church forests; we need less than $200,000 to complete our Noah's ark vision. We are focused on Ethiopia's northern region because that is where forest losses are greatest. My small TREE Foundation (www.treefoundation.org) serves as a central conduit for fundraising. But we need to act fast to save Ethiopia's biodiversity. We hope to attract either a large NGO to fund walls on a grander scale or a single philanthropist who would understand the importance of saving the species of an entire country. Stone walls are very inexpensive in comparison to saving tropical forests in other countries, and perhaps most important, the priests are trusted stakeholders and local communities seem happy to build gorgeous walls to safeguard their forests.

Our story is a win-win-win-win for conservation. The farmers win by removing stones from the fields to increase crop yield, the priests win by saving all of God's creatures, the people win because forests provide many services to enhance their lives, and the trees win because the walls ensure protection of both adults and seedlings. There is a strong belief that hermits, called *menagn* in Amharic, live silently under the canopy, invisible to most people but their presence is strongly felt. The chief priest, Abune Abraham, who himself lived for many years as a hermit, is now restoring an entire forest for his Orthodox diocese outside the city of Bahir Dar, to create a religious compound. He offered to build me a small stone hut so I can visit more often, a genuine gesture of good faith. By observing the priests who value silence, I realize my own childhood tendency toward shyness may not have been such a bad thing. I often think of my grandfather, who built our stone cottage, safely conserving a lone elm tree in the living room. I imagine he's smiling up in heaven, proud that I am building stone walls to save trees in Ethiopia.

⫸ Red Stinkwood or African Cherry ⫷
(*Prunus africana*)

DURING AN EARLY VISIT I MADE TO ETHIOPIA, the priest of the Debre Tabor church forest took my hand and led me into a nearby field where farmers were planting oats. In the middle of the mud-caked field stood one lone tree. Although we did not speak the same language, the message was noticeably clear. The priest pointed to this lone individual, an African cherry, and gestured to explain that it used to be inside the boundaries of his sacred grove. Determined to provide enough food for their families, the local farmers had overcleared, encroaching into spiritual grounds. The priest felt sad that many farmers were plowing too close to the church forest boundary trees, ring-barking trunks, which kills the edge trees in the process. Every time I return to Debre Tabor, that tree greets me as a sentinel amid a monoculture of crops, a stark reminder of the importance to conserve the remaining church forests.

African cherry (*Prunus africana*), also called red stinkwood because it emits odors, is distributed throughout the montane regions of central and southern Africa, as well as the islands of São Tomé, Grande Comore, and Madagascar. A member of the rose (Rosaceae) family, this canopy tree is the tallest in its genus, *Prunus*, reaching almost 150 feet in height. During his first expedition to the Cameroon ranges in 1861, the European botanist Gustav Mann and his team collected specimens that were shipped to Kew Gardens for classification. Back in London, the botanist William Jackson Hooker first named the plant *Pygeum africanum*, with the genus referring to its fruit shape, a depressed sphere resembling a human buttock. Nearly one hundred years later in 1965, the botanist Cornelis Kalkman made the case to move *Pygeum* to *Prunus* after using molecular cladistics to determine that this species was a cherry. Today, *Prunus africana* has many common names, including African cherry, pygeum, iron wood, red stinkwood, African plum, African prune, and bitter almond, as well as numerous names in many African languages.

African cherry has characteristic bark, blackish brown and fissured in a corrugated, rectangular pattern. The leaves are alternate, simple elliptical, bluntly pointed, glabrous, mildly serrated, and dark green above but pale green on the underside. The petiole is pink or red, and the plant flowers from October through May, with clusters of scented white or buff blossoms pollinated by insects. From September to November, reddish or purplish-brown fruits appear, similar to a cherry in shape and structure and eaten by birds, monkeys, and squirrels, as well as humans. Like other *Prunus* species, African cherry has extrafloral nectaries, glands along the leaf margin that provide specialized insects with a nutrient reward in exchange for protecting the foliage from herbivores. These plant-animal relationships are not well studied in most African trees, so we do not yet know which species frequent these canopies. In Rwanda, Dian Fossey reported that mountain gorillas love to eat the stinkwood fruits.

African cherry is highly sought for several medicinal uses and, as a result, would likely be overharvested if not legally protected under Appendix II of CITES (Convention on International Trade in Endangered Species). Its bark is used by local villages for many medical applications: malaria, wound dressing, arrow poison, stomach pain, kidney disease, gonorrhea, insanity, and fevers. The careless harvesting of bark has led to widespread tree mortality and subsequent decline of its distribution. Because of its reputation in treating prostate cancer, this species is now cultivated in many regions of Africa for pharmaceutical markets. A drug made from the bark extract was patented in 1966, and the retail value for this medicine is estimated at over $200 million annually. In addition to pharmaceuticals, African cherry is harvested for axe handles, wagons, floors, bridge decks, furniture, and utensils. Unlike its *Prunus* counterparts in Europe and North America, *P. africana* is not well studied and very few publications exist about any aspects of its ecology. Future botanical research is critical to adequately manage the fate of this important Ethiopian native, as well as just about every other African tree species.

11

CLASSROOMS IN THE SKY—FOR EVERYONE!
✴ ❀ ✴
Wheelchairs and Water Bears in the Treetops

T HE QUESTION I HEAR MORE OFTEN than any other from elementary schoolkids: "What is the most common species living in the canopy?" Unfortunately, there are not yet enough arbornauts to have figured the correct answer. But if I were to wager a guess, my response would be tardigrades, commonly called water bears or moss piglets. "Tardi-what?" most people ask. This relatively unknown phylum, Tardigrada, literally means "slow walker." These sluggish microscopic creatures don't really walk at all, but essentially float in a water droplet. They thrive in almost any moist substrate, fresh- and saltwater, so they can thrive in dry deserts with occasional downpours, moist tropical forests, and even the extremes of hot springs or Antarctica's icy cliffs. Any moist bit of moss, lichen, bark, or leaf surface provides the required film of water to coat their tiny cylindrical bodies plus four pairs of telescoping legs with claws or adhesive disks. And if their watery habitat evaporates, they transform into a dormant state to await rainfall, sometimes for decades, or they drift in the air above the treetops to a new location, seeking moisture. Neither drought nor flood nor extreme temperatures will kill them. When my kids were young, they learned about water bears from the popular television network Animal Planet, which dubbed them the most extremophile organisms on the planet due to their ability to withstand extraordinary physical conditions. About 0.2 to 0.5 millimeter in length (the size of a particle of dust), they dominate their Lilliputian kingdoms of soil, leaves, and water droplets along with other small creatures such as nematodes, collembola, rotifers, and mites. It sounds like a science-fiction invasion—billions of miniature bearlike creatures crawling across our suburban lawns and shrubbery while we sleep.

As a scientist, as an explorer—or really as any ordinary person living in this world—you never know what will bring you to your next discovery. Tardigrades came into my life because I was determined to provide opportunities in field biology for underserved youth. I have passionately pursued inclusivity in science for many decades, probably because I did not always feel welcome into my profession by the white-male network that dominated it. I was determined to ensure that young women and members of other minorities did not experience the same intimidation. At the age of eleven, I wrote my first book, *Through the Year with Gertrude Grosbeak*, about the escapades of a bird who befriended a bedridden child unable to walk. (But I did not publish it until age sixty!) One student who recently read and loved the book reminded me I had been pondering the challenges of inclusivity in field biology for almost a lifetime; as a wheelchair user herself, she could almost have been the role model for that book. In adulthood, I've made it part of my mission to take every opportunity to work with minorities in science—teaching girls from economically challenged families to climb trees at a summer camp, training female arbornauts donned in traditional garb in the jungles of India, hiring qualified candidates from minority backgrounds throughout every leadership position, sponsoring girls and boys from underserved communities for scholarships and research opportunities, and advising large numbers of the minority students at the universities where I taught. Looking back at those professorial days, maybe the label of single mom constantly juggling kids and work made me more approachable than other faculty members for students with similar challenges.

One group continually overlooked in field biology is mobility-limited students. Nearly 25 percent of Americans live with a disability, yet they comprise only 9 percent of the scientific workforce and 7 percent of PhDs in science, according to a 2019 report in *Science* magazine. The Americans with Disabilities Act (ADA) was launched in 1990 to ensure more inclusivity, but throughout my career, I had never encountered any program that encouraged students with physical disabilities to consider careers in field biology. Students who use wheelchairs are usually relegated to the laboratory or indoor office duty. As a botanical garden leader, I was frustrated by this, especially watching young people in wheelchairs, moms with strollers, and elderly people with walkers navigate the grounds with trepidation. It heightened my own awareness to realize they could not enjoy a forest canopy or probably even a bumpy woodland trail. That led to the building of North America's first skywalk accessible to people with physical disabilities. On the tenth anniversary of the ADA, our Florida representative, Pam Dorwarth, cut the ribbon from her wheelchair, officially opening the country's first ADA-compliant walkway in a botanical garden. Representative Dorwarth cried as a cluster of wheelchairs rolled up the gentle ramp to experience the treetops for the first time. But one ADA-friendly walkway was certainly not enough. I wanted to mentor mobility-limited students who dream of working outdoors as ecologists, who want to discover new species in forests. I felt certain that any student could climb if she/he had use of their arms to hang on to the ropes. I was never a star athlete and yet had earned a reputation as one of the world's most accomplished arbornauts. With advances in canopy access hardware, practically anyone can slide up a rope with minimal muscle power. My friend Patti had polio; she has since become executive director of Tree Climbers International, one of the world's largest recreational climbing groups. Both Pam and Patti are true inspirations to me, illustrating that tree climbing can be for everyone, and they were my trusted advisors in creating a canopy program for mobility-limited students.

What better way for students to become excited about field biology than discovering a new species? It is unlikely to find that new species of birds or beetles exist in temperate trees because those larger creatures have already been documented, but what about water bears? My son James was absolutely starstruck by water bears as a youth. During high school, he was determined to join a student expedition to Antarctica and study water bears living in the cold. First, he learned how to sample these tiny creatures from one of the world experts at North Carolina State University, who also shared his permit so James could legally collect critters and bring them back from a different continent. Second, he raised his own funds for travel because I simply couldn't afford such a major expense. He sold shares of stock in his water bear expedition to friends and neighbors who wanted to share the thrill of discovery. James not only sent a postcard to all his shareholders, but also hosted a scientific lecture upon his return to share his findings from the expedition. I guess it was in our family blood to work with water bears.

So that was how I ended up partnering with Randy Miller (aka Dr. Water Bear), a tardigrade taxonomist who himself has mobility limitations. The world of field biology is small, and the two of us first overlapped thirty years previously in Australia, where we shared the same lab and supervisor despite our different research projects. More recently, I reconnected with and funded him to come to the Malaysian BioBlitz, where he discovered new tardigrade species in the tropical forests. With a combination of my canopy skills plus his water bear expertise, we made a great team. Those are the unexpected and circuitous pathways by which collaborations in field biology are born. We never overlapped at a conference because our disciplines are so different, but we knew of each other's expertise and reputation, which formed the basis of a trusted partnership.

Randy hypothesized that water bears were common in temperate trees, but no one had ever surveyed them in the canopy, not even in Kansas, where Randy's home institution, Baker University, was located. We both thought that training student arbornauts to climb trees and collect tardigrades would yield new discoveries, so why not recruit mobility-limited students? We wrote a National Science Foundation grant to the panel called Research Experiences for Undergraduates (REU), which specifically targets college students in hopes of giving them exposure to scientific careers. After three years of us trying, NSF approved our grant. (Or maybe they just got tired of rereading the revised application?) Our project was officially entitled "3D Herbivory and Biodiversity of Tardigrades in North American Forest Canopies: Inspiring Students with Physical Disabilities to Pursue Field Biology," but our students affectionately called it "Wheelchairs and Water Bears in the Canopy." With tardigrades as our target phylum and the tree-tops as our fieldwork site, we surely had a slam-dunk program to guarantee a sense of exploration and achievement for our students.

Surprisingly, finding students who qualified for this grant was our first big challenge. There were no streamlined digital mailing lists of wheelchair users on American college campuses, so we advertised in various ADA publications and through scientific societies. The plan was to host teams of both mobility-limited and fully mobile undergraduates, so they would also learn to appreciate each other's strengths and weaknesses. At the advice of NSF, we limited our scope to immobility, because they did not think it would be easy for two professors to design safe field methods for a range of disabilities. Students with autism, blindness, deafness, or other limitations are equally deserving but require different methods to ensure a quality research experience. Funding student teams within the parameters of a research budget is always challenging; here, that included the costs of housing and flying students to Kansas, plus the required per diem stipends. Our small grant also needed to fund climbing equipment, laboratory materials, and the rental of vehicles that were wheelchair accessible. (Despite our efforts and several car companies' advertisements, the rental vans were not handicap accessible as advertised, so we lifted students in and out of their seats.) Baker University gave us a great deal for dormitory accommodation—$10 per night per student (compared to housing in my then work location of San Francisco at over $100 per night). So for budgetary reasons and because we knew we could find tardigrades in trees anywhere, the project made Kansas its base camp. I had almost forgotten how it felt to sleep on a plastic-covered dormitory mattress until I arrived at Baker University in Baldwin City, Kansas, on a sizzling humid June evening to begin the first of five summer research sessions studying water bears with undergraduates. Compared to other sleeping conditions such as remote jungles, it was luxury. I was also thrilled to have a flush toilet!

During the first week, we taught students to climb, employing my original, tried-and-true methods of single-rope techniques with harnesses, slingshots, and ropes. I'd used these techniques for over thirty years and knew they would work regardless of the students' mobility limitations. Still, there have been some improvements—cushioned harnesses, sleek aluminum slingshots (although intended for hunters, not arbornauts!), and climbing hardware carefully tested by recreational experts for safety. These all represented a huge advance from relying on homemade gear. Teaching wheelchair-bound students to ascend on a rope into the canopy was a pretty daunting prospect, but also one of the most rewarding experiences. We created a buddy system with pairs of mobility-limited and mobile students and recruited a Kansas City recreational climbing club, which provided two days of dedicated training sessions, so each student team had a highly trained arborist watching every move.

The first task was to learn knots and become familiar with ropes. We tied and retied knots, and then tied them again—Prusik, figure-eight stopper, clove hitch, and bowline were all good options that served many purposes and, as I always reminded the students, essential if you ever become marooned on a desert island! It was important to feel safe when dangling from a rope, and especially to feel confident your knots were correctly tied. Different hardware is connected via different knots; sometimes our sampling equipment was tied onto waist belts with knots or carabiners. The second challenge came with maneuvering the wheelchair tires over grass, soil, and roots to reach the base of a tree. The first practice sessions were held on the campus lawn, which offered fairly level ground. The third challenge, and probably the toughest action for students in wheelchairs, was to put on harnesses. If you are unable to stand up, it's not easy to slide each foot through stirrups and pull a waist belt up and around, but with our buddy system, everyone became fully rigged. The final and most exciting step was to clip an ascending device onto a rope, carefully affixed over a strong branch by our arborists (the students had enough challenges without rigging their own ropes during week one), and begin climbing, sliding up the rope like an inchworm. What a sight—three students launched simultaneously from their wheelchairs. Up, up, and away. Lots of grunting and perspiring, but we could see even from thirty feet away that every student felt a huge amount of pride and enthusiasm as they waved from a canopy perch down to the ground team. Over subsequent climbs, everyone became adept at sliding up the ropes and operating the jumars, the ascending gadgets with upward-facing teeth that allow ascent but prevent downward slide. Thanks to jumars, climbing a tree does not require strong muscles, but rather a fluid motion of sliding up a rope and moving with the hardware, a motion that ultimately comes to feel natural, like riding a bike. After the first day, everyone was extremely sore because it is impossible not to tense up and use muscles to grab the ropes or the trunk. But such physical force is not necessary, and ultimately, everyone learned to climb almost effortlessly. Over five summers, the training became even more effective as our teaching methodology improved. But there was always a lot of huffing and puffing, as is the case for any climber!

During week two, our training turned to the challenges of sampling water bears, not just getting into the trees. How do you find something you cannot see? Answer: you learn to predict where they are supposed to be and you collect there, hoping the water bears are floating blissfully within the sample! To successfully answer questions about the distribution of tardigrades in forests, our student teams created transects up entire trees and sampled moist substrates such as moss, lichen, foliage, and bark. In ecology, there is neither time nor necessity to harvest an entire population, just a small portion of it. It is critical to sample accurately and avoid bias, so much of our summer training involved demystifying field sample techniques and learning statistics. My prior experience researching leaves provided great preparation for water bears, because both required designing samples with respect to space and time. It was important to sample across spatial scales such as individual leaves, branches, heights, whole trees, and ultimately entire forests. Equally important was sampling with respect to time—days, months, years, and even centuries have an enormous impact on some field biological questions. In the case of water bears, no studies existed about any spatial or temporal variables: Did they inhabit specific canopy levels? Was there a preference for one tree species versus another? Moss, bark, or foliage? Because we could not see the microscopic critters, we randomly sampled all appropriate habitats, almost like making a best guess but keeping our fingers crossed. We collected up and down tree trunks, out on limbs with foliage and moss, and between different species. Each collected sample was about the size of a tablespoon of substrate, a mere tuft of bark or green stuff gently plucked and popped into a small paper bag carefully labeled with date, location, height, species, and habitat. Students learned to replicate each sample and to avoid bias, which might include something as simplistic as the temptation to consistently prefer the bright-green moss because it looks prettier than the grayish one. Over five summers, different teams of students sampled, sampled, and sampled—multiple collections for each height, species, forest, and time of day or night or season. As with leaf measurements, time and space were equally critical factors to answer questions about water bear distribution. Even more important, the good sampling regime saved time and effort, because the statistically correct design estimated numbers of tardigrades simply by subsampling a portion of the crown instead of counting every single water bear.

After mastering the nuances of fieldwork, we returned to the lab and learned to use microscopes, create slides, distinguish different species, and sort other creatures living alongside the water bears. Samples were air-dried in paper bags and later moistened with distilled water in a small container with a crosshatched bottom for counting. Students started their water bear hunts using a dissecting microscope at 40X magnification, so floating animals could easily be spotted. That first day, there were lots of shrieks and exclamations of "I got one!" as students trained their eyes to locate these pudgy floating blobs under the magnification. Next, specimens were transferred onto a glass slide for safekeeping and later identified to genus using a light microscope on higher powers of 200 to 400X. In some cases, students traveled to the

University of Kansas and photographed specimens with an electron microscope, which really enhanced their identification features for publication. After many episodes of climbing and laboratory work, we published the results. And publish we did—more than fifteen scientific papers featuring over twenty-five thousand water bears collected, plus eight new species discovered to date from common temperate trees. Our grant became a poster-child project for the National Science Foundation in terms of outcomes as well as inclusivity. Student teams had extraordinary diversity—disabled, fully mobile, community college attendees, military veterans, moms with kids, Black, White, Latino, Asian, and first-generation college students.

Despite their ubiquity, water bears are one of the least known groups of organisms living in everyone's backyard. Tardigrades are related to insects, but different enough to earn their own phylum on the evolutionary tree between the nematodes (roundworms) and the arthropods (crustacea, insects, ticks, and mites). Over one thousand species of water bears have been described in the scientific literature, and they exist on every continent, but because there are so few global experts, they were never studied in canopies until now. Like higher animals, they have digestive, excretory, musculature, and nervous systems. But like lower animals, they lack respiratory and circulatory systems, instead breathing through their skin. Some species are vegetarian and others are carnivorous, feeding on bacteria, algae, tiny soil organisms, and sometimes each other. By clinging to leaf surfaces, tardigrades are in turn eaten by vegetarians, such as deer and cattle. They are most certainly eaten by unsuspecting humans as tiny morsels of nutrition in salads or vegetables but are certainly not visible to the naked eye. Under the microscope, tardigrades look like little bears, and I must admit, they are really cute! Their unique habit of survival in a suspended state may someday prove important for medical research or for space travel.

Technically, Animal Planet television was not exactly correct to call them extremophiles, because tardigrades are classified as "extremotolerant" organisms, which is the secret of their capacity to endure environmental adversity. During droughts, they enter a dormant phase called cryptobiosis, transforming into a tiny desiccated ball called a tun. During other environmental extremes, they survive by swelling up like a balloon and floating into the atmosphere, scientifically termed anoxybiosis or commonly known as tardigrade rain, to seek better conditions for their lifestyle. Tardigrades have come back to life when a drop of water was administered to a tun in a dried plant collection over a hundred years old! Another was revived from an Antarctic moss sample that had been stored and frozen for over thirty years. They have flown as experimental organisms on NASA expeditions and not only survived, but also reproduced in outer space. During 2019, the Israeli *Beresheet* probe was intended to be the first private vehicle to land on the moon, until the robotic craft lost contact with mission control and crashed. Organizers believe its cargo, including a few thousand tiny water bears, may have survived when the draft released its payload onto the landscape. This means a population of tardigrade tuns could now exist on the moon, a harsh environment for mammals but entirely tolerable for the toughest animal on Earth! Technically, they need an atmosphere plus water to colonize, but can exist as tuns for a long time. Will scientists find water on the moon? Who knows, but if it does exist somewhere on the lunar landscape, water bears may be the first to detect it.

Our students addressed questions new to science: What is the density of water bears in different canopies? Are there hundreds per tree, or millions? Of the same species, or different? Over five summers, we collected 28,384 samples representing 37 species from 58 forests in 4 states, ranging from zero to 4,500 feet in elevation and from zero to 200-foot-high trees. Our students climbed 492 crowns and discovered 8 new species, plus set 26 new distribution records. What was especially amazing (called an "OH WOW factor" in field biology) is that 80 percent of every sample of lichen, moss, bark, or foliage collected contained at least one water bear! And it is likely the other 20 percent also had a water bear, but our microscopic techniques were not so exacting with amateur eyes. Because of the labor-intensive process to analyze each sample, most of our questions posed remained unanswered. It will probably take years to analyze enough samples to determine how many tardigrades, on average, live in the canopy. But based on preliminary findings, there are probably over a billion tardigrades per ten acres of forest. Over five summers, we focused on the local trees of Kansas, because it was most cost-effective with our limited budget in the research grant. But each summer we also visited one field station or national forest around the country, asking a curious question: Does water bear density differ in forests across the country, such as between oak-maple stands in Massachusetts and old-growth conifers in Oregon? To answer that, we visited several ecological field stations across America to collect samples. Preliminary counts indicate that Massachusetts temperate forests have more water bears than Kansas, Florida, or Oregon, but as with all ecological research, one summer of sampling is not conclusive. One finding emerged from all sites, including Kansas: higher densities of tardigrades live in the treetops than the understory. As with so many aspects of field biology, especially biodiversity, both funds and labor are limited, so the collections are slow to process.

Our most daunting sampling escapade involved climbing in the Pacific Northwest, where the average height of a mature western conifer is about two hundred feet. Our students were already adept at climbing fifty-foot-high oak and ash in Kansas, but four times more distance to climb is a lot of perspiration. To ensure absolute safety, we recruited an arborist who was particularly qualified to rig tall trees. After flying from Kansas to San Francisco, California, our western expedition started with a walk on an ADA-accessible trail under the coastal redwoods (*Sequoia sempervirens*) in Muir Woods. That evening, I asked the students to write in their journals about how they felt to first set eyes on some of the world's tallest trees. One commented, "Looking at the size of these trees and the magnitude of their trunks makes me realize how small I am in the world, and especially in the universe." How true. The Pacific Northwest is home to the world's tallest species, yet fewer than 10 percent of Americans have ever seen a redwood. More people have climbed Mount Everest than scaled a redwood! Despite their stature and geographic location in one of the most science-savvy regions of the world, almost nothing is known about their canopies. It is not easy to scale these giants, with such enormous height and lack of side branches for positioning a rope. Other tall denizens of Pacific Northwest conifer stands include giant sequoia (*Sequoia gigantea*), western hemlock (*Tsuga heterophylla*), Douglas fir (*Pseudotsuga menziesii*), western red cedar (*Thuja plicata*), noble fir (*Abies procera*), and Pacific silver fir (*Abies amabilis*), all equally daunting to climb. With the advent of high-powered slingshots and extra comfortable harnesses, forest scientists can now safely explore the upper reaches of the crowns. Our students sampled in several different crowns and a week later, back in the lab, discovered that these moist canopies represented ideal habitat for water bears, with one dominant tardigrade, *Pilatoibus oculatus*, comprising 52 percent of their collections. Redwoods often live surrounded by fog, and that moisture undoubtedly offers a welcoming ecosystem for water bears. Their crowns not only produce energy from sunlight in the conventional fashion by drawing water up from their roots, but occasionally shortcut the system by directly absorbing fog into their foliage through stomata. In the past decade, the botanist Todd Dawson at University of California, Berkeley, and his students discovered this unique pathway where fog offers an advantage for photosynthesis, explaining why these trees thrive in foggy coastal areas along the Pacific coast. In addition to providing water directly to the needles, fog surrounds the crowns and reduces evaporation, increasing the water efficiency of redwoods and other neighboring vegetation. These tall trees not only directly absorb water into their foliage via fog, but also take in moisture in the conventional fashion, from the roots to the leaves via a vast network of xylem cells. Like a straw in a milkshake, the xylem tissue operates with a suction mechanism that draws water upward as a key ingredient for photosynthesis in the foliage. Water moving up a tall redwood represents one heck of a big straw!

Current climate-change models predict decreases in fog for California's coast as well as the advent of more frequent droughts. The Arizona State University biologist Greg Asner conducted aerial surveys of forests before and after the extreme droughts of the years 2014 through 2016, working in collaboration with the researchers Anthony Ambrose and Wendy Baxter, who climbed the trees and provided ground-truthing by confirming the results of the imagery with an up-close examination. By repeated canopy measurements from within the canopy and from afar, Greg, Anthony, and Wendy found that individual canopies transpired upward of five hundred to eight hundred liters (132 to 211 gallons) of water per day! This enormous figure far exceeded estimates based on ground-based calculations alone. Since then, teams in airplanes have surveyed Northern California forests to monitor the impacts of drought, assessing losses in water content using laser-guided spectroscopy and satellite-based models. The results indicated an astounding number of dead crowns, and severe water loss in over 30 percent of a one-million-hectare forest, comprising approximately fifty-eight million large trees. As climate change increasingly causes extremes in both rainfall and fog, such moisture fluctuations represent an enormous threat to the future health of these otherwise resilient giants.

For scientists like me who study herbivory, the redwoods are puzzling because they are almost devoid of foliage feeders, despite the enormity of their salad bar. In fact, they have escaped predation over millennia, which is almost entirely unique in the world of ecology where at least one insect usually evolves to defoliate every species of plant. Paul Fine at the University of California, Berkeley, is currently studying the toxins of redwood foliage and bark, to figure out how they have remained immune to insect attack over thousands of years. A few historic observations have cited very mild impacts from cone moths and roundheaded borers attacking cones and seeds; foliage nibbled or sucked by aphids, scales, mealybugs, leaf beetles, and a few others; twigs attacked by bark beetles and twig borers; buds nibbled by tip moths; and occasional bark stripping by bears. Unfortunately, in 2020 the first mortality of a giant sequoia from insect attack was reported in Sequoia National Park, California. This tree was over two thousand years old, and its neighbors had recently died from drought and fires, leaving the remaining individuals more vulnerable. Bark beetles had invaded the upper branches and caused the first documented death of a sequoia from insect attack, presumably because the trees were exceptionally vulnerable due to climate extremes. Sadly, similar deaths are occurring with other conifers throughout the region. As Dr. Christy Brigham, who oversees the forests of Sequoia and Kings Canyon National Parks, admitted to *The Guardian*, "This is not how giant sequoias die. It's supposed to stand there for another 500 years." As climate change becomes more extreme, it is beginning to impact even the world's biggest, longest-lived giants.

It is not easy to study the canopies of the world's tallest trees. Not only redwoods in California, but the dipterocarps in Malaysia, great kapoks of the Amazon, and even the cardboard trees of Africa posed great challenges for safety and scientific accuracy. Ropes and sling-shots are not effective for their upper reaches, and walkways are not always feasible to engineer. A fourth unique tool in our arbornaut's toolkit that has been utilized for the highest crowns, including in the Pacific Northwest, is a construction crane. Albeit expensive to operate because it requires a highly paid professional driver, the crane provides easily repeated access to the uppermost reaches of the highest crowns, because its crane arm rises above. One crane requires about a million dollars to acquire and install (usually purchased secondhand from a construction company), in addition to operating expenses. As a reality check, that cost is minuscule compared to the budgets of NASA or a particle accelerator, which require billions of dollars, but the budgets for field biology are significantly less than for outer space exploration or physics. The Wind River Canopy Crane was operated in the Pacific Northwest by the University of Washington during the late twentieth century, allowing unique access to some of the world's tallest trees. I was fortunate to be one of a handful of arbornauts who used this crane for research. Our team sampled insect damage in the tops of Douglas fir, western hemlock, Pacific yew, and western red cedar. We created a sophisticated and highly accurate sampling regime, generating 101 random points in the canopy with a three-dimensional computer model, and then used the crane bucket to access each sampling point over the period of one week and conduct replicated measurements of insect damage to the foliage. The results showed the least insect damage of any forest type in the world, with an average of 0.3 percent leaf area consumption. This is next to nothing, especially as compared to tropical foliage in Australia and the Amazon averaging 15 to 30 percent leaf area consumption. Half of the sample points had no detectable herbivory on the conifer needles, making this one of the world's most resistant foliage. These conifers contain effective defense chemicals and appear to stay ahead of the rapid adaptation of insects to digest such toxins. Because one old-growth Douglas fir contains well over one million leaves, accurate subsampling to measure herbivory was key to obtaining accurate results.

That same Oregon crane bucket was also ideal for sharing the treetops with middle school students who could never have climbed up 150 feet on their first attempt. As part of the JASON Project virtual expedition series, I took approximately ten sixth graders on a mini expedition into the enormous Wind River crane bucket to survey some of the world's highest foliage. One youth from the Bronx gasped with excitement when he saw a banana slug on a branch as our gondola car rose higher, and he grinned from ear to ear. Cranes made it easy to share the eighth continent with even the most landlocked citizens, or with the least likely arbornauts. Approximately ten canopy cranes exist around the world, including two in Panama, several in China, one in Australia, some in European temperate forests, and a few in planning stages. Unfortunately, the Pacific Northwest crane has ceased operation after funding cuts and a history of political skirmishes between loggers and environmentalists. Now China leads the way with three cranes, and Germany boasts two, all dedicated to canopy research.

I had my first experience operating a crane when I went to survey herbivory in the tropical forests of Panama. How exhilarating to stand in the bucket up close to a large cluster of iguanas sunbathing in the upper branches of *Swartzia simplex* (later calculated as 10.1 percent herbivory), and to sample hundreds of sun leaves in one day using the comfort of a four-foot-by-four-foot crane bucket to maneuver between crowns. Compared to single-rope techniques, conducting research from a crane bucket is easy; the toughest challenge in Panama for me was learning to communicate directional vocabulary in Spanish to a crane operator via walkie-talkie. As with most leaf research, I am a stickler for not bumping the branches or tearing my subjects—so I kept reminding the driver to steer delicately. Cranes require almost no perspiration, but they are too expensive for most field teams to afford, and sampling is limited to the reach of the crane arm. Even walkways offer more flexible canopy access because bridges and platforms can be moved or expanded at a relatively low cost, plus offer access twenty-four hours per day without the expense of hiring professional drivers. But cranes have inspired canopy research in the Pacific Northwest as well as in Panama and Germany.

Although we did not use the canopy crane for collecting, our water bear findings from the Pacific Northwest are the first records of this phylum from these tall trees, but they may require a decade to fully analyze. Regardless of the slow process of classifying and publishing the findings, our mobility-limited students can be proud of their achievements, and I hope more than one of them will pursue a career in field biology! A star student named Rebecca, who was shy and reminded me of myself, later accompanied me to the Amazon jungles, where she undertook research in tropical trees. Loading and unloading her wheelchair from small dugout canoes was not easy, but the trip fulfilled a lifelong dream for her to experience the Amazonian rain forests—and she is now a seasoned arbornaut! Not bad for a reticent young lady with limited mobility working alongside her treetop supervisor, who still sometimes acts like a wallflower, even in adulthood.

» Coastal Redwood «
(*Sequoia sempervirens*)

A FRANCISCAN MISSIONARY, FRAY JUAN CRESPÍ, authored the first written record of redwood trees in his diary on October 10, 1769, near Monterey Bay, California. Imagine those early explorers struggling to advance their brave teams through the treacherous Northern California terrain of ancient Paleozoic and Mesozoic rocky slopes, only to enter a valley of enormous reddish trunks. Crespí was the only Franciscan monk to traipse from Baja California northward to what is now San Francisco, as the official expedition diarist, and he wrote of "very high trees of a red color, not known to us . . . In this region there is a great abundance of these trees and because none of the expedition recognizes them, they are named redwood for their color." From his simple descriptive journal entry, a humble missionary gave rise to the name of America's tallest and most iconic species.

In geological nomenclature, redwoods are called "paleoendemics," meaning a species with an extant range that represents a remnant of its former distribution in the fossil record. This species is, in a sense, a living fossil. First recorded among the fossils of the Mesozoic Era some 150 million to 200 million years ago, redwoods had a nearly circumpolar arctic distribution as well as extensive mid-latitude range and were common throughout the western United States, Canada, Europe, Greenland, and China. As the climate became cooler and drier almost two million years ago during the Quaternary, remaining stands shrank to their current distribution along the Pacific coast of North America. Extensive logging practices during the nineteenth century reduced these fragments even further to 120,000 precious acres, according to Save the Redwoods League at the time of their centennial in 2019.

Early redwood science was primarily based around extracting timber. Many forestry reports cited estimates of wood volume for different old-growth stands, plus their variation in structure and species composition. In 1934, the forester W. Hallin published the largest volume of wood per unit area; he recorded 10,856 cubic yards per hectare, which represented 178 stems. (Based on the average home requirement of 2,480 board feet in 2011, that is enough timber for 188 homes!) Other twentieth-century research included the interface of stand dynamics and environmental conditions, as biologists sought to understand how the trees grew so tall. Despite an overarching interest in harvesting these giants, biological curiosity also grew, along with a strong conservation ethic. Only in the last few decades have microclimate, photosynthesis, soil mats on branches, light requirements, and canopy biodiversity become hot topics, but initial research was predominantly limited to ground-based observations.

One of the newest discoveries about redwoods involves their water hydraulics. Most foliage absorbs carbon dioxide in leaf surfaces through openings called stomata, and in turn releases water in a process called transpiration. This triggers a siphon (or straw-like) train of water molecules pulled up from the roots through the xylem cells, a virtual water channel up the trunk. In short, trees serve as the exhale to our inhale, filtering carbon dioxide from the air and exchanging it for oxygen. This photosynthetic process whereby trees can produce energy from the sun using hydrogen and oxygen from water is especially productive in redwoods, because their height allows for so many layers of foliage to operate simultaneously. Leaf area index has been measured at 14.2, meaning each square yard of forest floor contains 14.2 layers of foliage overhead. If one were to dangle a string down from the top of the canopy, it would intercept 14.2 needles! That makes for a lot of photosynthesis going on overhead! (Contrast this statistic with New England deciduous forests, which have a leaf area index of approximately 4 to 6, meaning a string dangling through their canopy depth would intersect only 4 to 6 leaves.) Other creative water strategies in these crowns include the direct intake from fog into leaf stomata, as described in the previous chapter.

The science of redwoods advanced significantly after the advent of canopy access. Using new technologies for arboreal ascent, the Humboldt State University botanist Steve Sillett and colleagues were the first to document biodiversity in the upper reaches of redwoods. They found 282 species of epiphytes in 9 crowns, including 183 lichens, 50 bryophytes (mosses), and 49 vascular plants. For a temperate forest, this was record-breaking! Their tenacity to climb over two hundred feet to the tops of crowns not only represented a lesson in courage, but also a new chapter in forest ecology. Sillett applied his climbing prowess to study a big arboreal mystery: How can water get to the top of such a tall tree? He was the first to measure the incredible straw-like action of water molecules traveling from the root hairs up through the small network of xylem cells called tracheids, all the way to the needles some hundreds of feet aboveground. Even more incredible is that water moves quietly through trees—no one ever hears the gurgling of their fine-tuned machinery while walking through the woods. Part of the efficiency of water transport is because the uppermost foliage is smaller and thicker than its understory counterparts, and this efficient shape not only survives in high winds and storms, but also endures the enormous tension created by the upward flow from roots to crowns. Such extraordinary tension of water up a trunk exerted by its leaves is called capillary action. Scientists have shown that mature individuals, with such efficient waterworks, grow more wood than younger ones, making this species an enormous storage unit of carbon sequestration, which continues to expand as it grows older. This is an opposite pattern from mammal species, which decrease their growth rates upon maturity. It is perhaps no surprise, because older trees retain more foliage than younger ones, allowing them to grow bigger and better with age, transforming ever-expanding amounts of carbon dioxide into wood production. And one last factoid discovered by arbornauts relates to the complex crowns of redwoods: the trees sprout new leaders (shoots at the branch tips) in the upper canopy after wind or storms have damaged the existing trunks. Such response to repeated weather conditions results in complicated masses of separate leaders, both living and dead, and massive amounts of detritus in the tree crotches where entire mini communities establish. One individual tree called Ilúvatar, named after a character from J.R.R. Tolkien's *The Hobbit* and *The Lord of the Rings*, contains 220 different trunks branching in the crown, representing regrowth from fire or wind and comprising over 37,000 cubic yards of wood. Measured by Steve Sillett and colleagues, this tree is considered the most complex living organism on the planet, but only by climbing into its upper reaches were such structural wonders discovered.

Not only are redwoods physically complex, but their genetics are similarly extraordinary. Recent research indicates they have a whopping sixty-six chromosomes, compared to most conifers with only twenty to twenty-four. In comparison, our human genome contains a mere forty-six chromosomes. This anomaly, called hexaploidy because this species has six copies of each chromosome, explains how they are capable of such enormous genetic variation. But just over one hundred thousand precious acres of primary stands remain in the Pacific Northwest, representing less than 5 percent of their original distribution before human activities destroyed these giants. Scientists recognize that climate change, with its predicted decline in fog and more frequent droughts, will further endanger redwoods and other Pacific Northwest conifers. Save the Redwoods League funds much of the ongoing canopy exploration, providing new insights into the trees' survival. Coastal redwoods are not the world's largest trees—this title belongs to their inland cousin, the giant sequoia. But they are the tallest, reaching 379 feet high. Were one to sprout next to the Statue of Liberty, it would be capable of growing taller than her extended torch. And such growth might require over two thousand years, according to dendrochronologists who study tree rings.

Can canopy science save the redwoods? No, but it can help by bringing attention to the value of these magnificent trees. Perhaps we humans can learn from ancient giants. Maybe they can teach us lessons about adaptability and survival. These trees exhibit both resistance and resilience, which are increasingly important attributes in a changing environment. Redwoods are the tigers of the botanical kingdom—an iconic species that exemplifies endurance and inspires a sense of wonder for the natural world.

12

CAN WE SAVE OUR LAST, BEST FORESTS?
⫸⫷

Promoting Conservation Through Mission Green

Destroying a tropical rain forest and other species-rich ecosystems for
profit is like burning all the paintings of the Louvre to cook dinner.
—E. O. WILSON, emeritus biologist, Harvard University

IN 2020, AFTER TRAGIC FIRES BURNED MILLIONS of acres in the
Amazon, Australia, Indonesia, California, and other landscapes, I
was interviewed by the BBC. They only asked one question: What
will happen if all the world's forests disappear? My answer was simple but stark: People will not survive. Period! Trees provide essential
functions that keep us alive, and without those green machines with
their zillions of efficient energy factories (aka leaves), no life on Earth
can exist. As an arbornaut, I shout ceaselessly about the importance of
trees and how they keep the planet healthy as well as all of humankind. Especially big trees! Here is a simplified list of ten ecosystem
services that forests provide, even as we sleep:

1. Fresh water
2. Climate control
3. Medicines

4. Building materials
5. Carbon storage
6. Energy production
7. Food
8. Genetic library of millions of species
9. Soil conservation
10. Spiritual place

People can't survive on a planet where the loss of tree roots allows soil to erode into our waterways, where foliage no longer cleans and circulates fresh water once canopies are gone, where over half of terrestrial biodiversity has nowhere to live, where our biggest carbon storage warehouses have been clear-cut, or where we have lost tree cover that shades, cools, and shelters us. After the devastating fires of 2020 in Australia, our family farm is gone, after six generations. An estimated one billion animals were consumed in flames, victims of a record hot, dry spell in the former "lucky country," which scientists linked directly to human activities warming our planet. After such fires, it will take more than a few seasons of planting seedlings to restore the Australian ecosystems, and probably more like a hundred years to restore enough canopy cover to support the koalas, currawongs, and kangaroos.

The Royal Statistical Society's decadal competition, an award for the 2010–2019 global statistic that spotlights the world's most pressing issue, revealed a stark harbinger of the fate of global forests: their dubious honor went to "8.4 million soccer fields of land deforestation in the Amazon over the past decade" (also equivalent to 10.3 million American football fields or twenty-four thousand square miles). Despite the well-intended efforts of the Royal Statistical Society to highlight this horror, the pace of clearing is accelerating. All three major primary tropical rain forests of Southeast Asia, the Amazon, and the Congo Basin are rapidly shrinking due to human activities. In addition, major woodlands in the temperate regions of northern Russia, China, and Canada have also experienced recent catastrophic fires and excessive clearing. It is not just about acreage, but also about biodiversity and tree age, girth, and height; ultimately, the large tracts of primary or old-growth forests are the most precious. I will say it again: save big tree canopies! Our planet houses approximately 60,065 species of trees, with over half found in just one country, meaning they are endemic and at risk of extinction if that region is deforested. Plants have more biomass than any other kingdom of life, tipping the scales at 450 gigatons (GT) of carbon, as compared to a mere 1 GT for arthropods, 0.06 for humans, and only 0.002 for wild birds. (A sobering fact is that wild mammals weigh in at 2 GTs, but domesticated livestock twenty-twofold higher.) Total plant biomass has declined twofold to 450 GTs since humans dominated the planet as forests are increasingly cleared. Foliage absorbs the excess carbon dioxide that humans pollute into the atmosphere, and wood is primarily stored carbon, so big trees represent excellent carbon sinks. Our remaining fragments of tropical rain forests are storing less carbon now than twenty years ago, in part because we remove the old growth and, at best, replant seedlings in hot, dry cleared landscapes. In the 1990s, Amazon forests removed forty-six billion tons of carbon dioxide, but in the 2010s, this figure declined to twenty-five billion tons. With the continued human impacts of climate change causing fire, drought, heat, and insect outbreaks on those remaining stands, climate modelers now predict the Amazon could become a carbon source by 2035 rather than a carbon sink, given its current trajectory of degradation.

Those same big trees not only store carbon but also shade the soil and moderate global temperatures. A study by Brazilian scientists in the journal *PLOS One* (2020) revealed that when different amounts of Brazil's Atlantic rain forest were experimentally cleared, the air temperature rose as much as four degrees centigrade after total clearing and one degree centigrade from 25 percent clearing. Research in subtropical forests of China showed those with higher diversity were better able to combat drought than stands with low diversity. With an estimated 66 percent of humans expected to live in cities by 2050, urban trees also deserve priority as a valued commodity. The US Forest Service calculated economic values of canopy cover in downtown Austin, Texas, at $34 million annually for the trees' ecosystem services. But the average lifespan of an urban tree is shortening, usually cut down after a mere nine years of life to make way for the onslaught of wider roads or new construction.

Tropical forests not only serve as Climate Control Central for our entire planet, impacting gas exchange and rainfall patterns quietly and efficiently, but they also house the greatest amount of biodiversity, with an estimated two-thirds of terrestrial species living on 10 percent of Earth's land surface. Most of my career has been dedicated to exploring biodiversity in tropical treetops, with the extraordinary estimates by arbornauts that 50 percent of our terrestrial species inhabit the upper reaches of trees yet approximately 90 percent remain unclassified by science. In short, we are destroying species before we even discover them, including plants that convey resistance to disease, unique pollinators for figs, or certain canopies best adapted to intermittent heat and drought. In his book *Half-Earth*, the renowned Harvard biologist E. O. Wilson makes a case for saving half of our planet to house 99 percent of biodiversity and the other half for one species, *Homo sapiens*. Wilson lists seventeen critical forests of highest importance to prioritize for conservation, including my own research priorities: the church forests of Ethiopia, the Western Ghats of India, California's redwoods, Southeast Asia, and the Amazon basin. Using his well-researched list as a blueprint, I have launched a new project called Mission Green to prioritize saving Earth's highest-biodiversity canopies. Endorsed by my colleague the distinguished oceanographer Sylvia Earle, who operates Mission Blue, I will mirror her goal of finding "hope spots" in oceans, which she defines as areas of healthy waters and high diversity. Instead, I will identify "hot spots" of biodiversity in global canopies and build walkways to provide an economic incentive for local people to earn sustainable income from ecotourism instead of logging. Mission Green is almost half executed and funded, with operational skywalks in Malaysia, Florida, Amazonian Peru, and Rwanda, and new ones underway in Mozambique and the California redwoods. Other forests urgently need walkways to inspire conservation before too much degradation occurs: in Madagascar, India's Western Ghats, Papua New Guinea, and the Congo. After creating walkways that can employ indigenous people (especially women), my small foundation (www.treefoundation.org) hopes to fundraise for student scholarships to study biodiversity at these canopy hot spots and continue documenting those 90 percent unknown creatures living above our heads.

The only silver lining of these recent fires and excessive deforestation is that such disasters have highlighted the global acreage and destruction of trees in stark numbers, inspiring millions of citizens to get muddy and help restore the planet. Ethiopia made the *Guinness Book of World Records* by planting over 352 million seedlings in 2019 during a twelve-hour span. Less than one month later, the region of Uttar Pradesh in India planted 220 million seedlings in one day—approximately one for every resident of that state. Citizens of both countries participated in these massive plantings by digging holes and watering seedlings. The Swiss Federal Institute of Technology published a report claiming that the planet could support an extra 2.2 billion acres of tree cover, using lands currently underutilized or areas where forest was cleared. They further calculated that after several decades of growth, these newly planted landscapes could remove about two-thirds of an estimated 330 billion tons of carbon spewed into the atmosphere by humans since the Industrial Revolution. Similar aerial imagery was used to count all the trees on our planet, estimating that Earth still houses some 3 trillion (that is, 3,000,000,000,000!) with Russia holding approximately 642 billion trunks, followed by Canada and Brazil, and the United States coming in fourth with 228 billion. Although some scientists dispute the exact numbers reported in these studies and planting events, canopy cover is broadly recognized as essential to the health of our planet. Saving big trees remains the best option for stewardship of natural resources, with planting seedlings a distant second choice. After all, young plants face extraordinary odds to survive. They confront the biological hurdles of competition, trampling, obtaining enough sunlight and water, and predators, as well as the additional human-induced threats of fire, drought, rising global temperatures, and large-scale clearing. The chances for even the most robust seedlings to reach adulthood are daunting, and if they manage to grow into adults over many decades, only then can they house biodiversity, store tons of carbon, and create a healthy forest ecosystem.

And that is a big "if" because much of the original biodiversity may already be extinct.

We have a lot of forest, and are planting more, but over fifteen billion trees are cut down each year. (Numbers of trees planted are not comparable—they vary drastically year to year, and they do not come close to equating the ecological and economic value of mature trees.) An estimated half of the world's primary (meaning original or old-growth) forests have been completely destroyed since most baby boomers were born. Not surprisingly, those old-growth stands housed most of the global carbon, significantly more than what is stored by seedlings struggling to survive in the harsh soils of Ethiopia or India. It is not okay to cut down tall trees and then plant a few small ones in exchange. We need to shift our response to climate change away from a quick fix of finding simplistic ways to store carbon to creating long-term solutions that reduce carbon pollution. I will repeat it again: big trees are an enormous global asset. Two centuries ago, the United States cleared over 95 percent of her original forests, but temperate landscapes are easier to restore than tropical ones, and the burgeoning American economy could easily afford the cost of reforestation. But that is not the case for Brazil, Madagascar, and the Congo, where it will require hundreds of years and significantly more expense to fully restore a complex tropical canopy, as compared to a temperate European or North American woodland.

So, what can be done to conserve forests, now recognized as one of our best planetary assets?

First, we must make sure all humans have a chance to encounter the awe-inspiring wonders of trees. I encourage parents to take their kids on a canopy walkway, not just a roller coaster. When families and citizens experience the canopy, they come away educated about the incredible complexity and magic of this green treasure, and more motivated to protect it.

Second, we must be mindful of our spending and how we (perhaps inadvertently) contribute to the loss of trees. We must realign the buying power of industrialized countries. Deforestation in the Amazon is largely driven by consumers in the temperate zones who buy tropical timber (often illegally imported), soy, tropical fruits, beef, and palm oil. If individuals with the greatest purchasing power would insist their governments label products with the geographical origin of their coffee or soy, then our wallets could determine successful conservation outcomes. Please ask for shade-grown coffee, harvested under the rain forest canopy, not coffee cultivated in open sunshine after clear-cutting. Insist on buying products without palm oil (and beware because some manufacturers use some twenty-plus other names for palm oil to fool consumers). Do not buy timber, soy, or beef from tropical countries, and ask your grocery store to insist upon accurate labeling from their supply chains. Write to your politicians to request labels of the energy footprint required to get a product into your hands as well as its geographic origin.

A third thoughtful action is to become a citizen scientist. Join an expedition with Earthwatch, a local BioBlitz, or an urban canopy cover survey to contribute to our knowledge of trees. Field biologists like me welcome the help of the public to count bugs or measure leaves, and more eyes and hands will achieve better results. Why not host a local BioBlitz in your own backyard or nearby park, joining with other families? Kids will love it, and perhaps your city government can use the data, as well as your local arbornauts. And fourth, read all you can about forests and share your knowledge with family, Sunday school, sports teams, community groups, friends, and teachers. Why not become an amateur expert about dendrology and help others learn? Without knowledge, people are less motivated to conserve forests.

Fifth, become a steward of big trees, especially since (as Dr. Seuss says in the book *The Lorax*) trees have no voice. Advocate for them in your community, and do not allow malls, buildings, or road construction to assassinate your senior green giants, making them victims of "progress." If people could build the pyramids or Machu Picchu many hundreds of years ago, then today's developers can surely utilize our advanced technologies to retain a few big trees amid our epidemics of cement. Planting seedlings or even saplings is great, but nothing takes the place of a large, mature canopy in terms of filtering water and producing oxygen. And all the creatures in your community will be so grateful you are safeguarding their home. Big trees are a life insurance policy for the next generation.

In this book, I depicted the life of a scientist, more specifically a field biologist who studies biodiversity, starting with that critical childhood pastime of playing outdoors. The conventional recipe for a field biologist: Read. Observe. Ask questions. Keep a journal. Conduct field research. Follow up with laboratory experiments. Sample with statistical accuracy. Collect enormous databases. Publish. Revise. Resubmit manuscripts. Design experiments to perfection, controlling every outside variable. Publish. Write grants. Rewrite grants. Collect more large data sets. Compete for tenure. Write more grants. Collect more data. "Publish or perish" has long been considered the mantra of a successful career scientist. But mid-career, my trajectory changed to reflect a new infusion of actions, urgently needed if we are going to save the very forests that we scientists are studying: Educate kids about trees. Explain ecology to the public. Engage diverse stakeholders. Write for mainstream media, not just technical journals. Seek direct actions to conserve forests. Shout about it. Train the next generation of conservation biologists. Engage women as environmental stewards. Forge alliances with priests, corporate leaders, and other community stakeholders. Share stories about ecosystems. Mentor girls in science. Collect data. Partner with citizen scientists. Build walkways to save forests. Write about it. Study forests where no one has funded prior research. Give tree books to indigenous kids. Teach students to become future arbornauts. Talk about it. Share ideas with anyone who will listen. In the end, my grandchildren may not read or even appreciate my arsenal of technical publications or my twenty-five-page resume, but I fervently hope they appreciate the native forests of Ethiopia or marvel at the view from a canopy walkway in Florida, Vermont, Australia, or Malaysia.

I have dedicated almost six decades to a lifelong passion for plants. I was a fan of pistils not pistols, and beetles not Beatles. I have shared my story of leaves and slingshots here in hope that it might resonate with a young person who is labeled as a nature nut because she prefers to hear a robin sing instead of play video games. Maybe my misadventures will encourage readers of all ages to look differently at a treetop and appreciate the complex protocols of how field biologists study them. Even as an adult, I am labeled as the eccentric client in the local bank who, when a hurricane is bearing down, brings her bugs-in-vials to the vault while everyone else is lined up to safeguard diamond jewelry. My children tease me to switch titles from arbornaut to arbornut, because the world of forest conservation is so downright depressing and riotous. But never underestimate the power of one. I am only one arbornut, and I can only do so much. But I do know the world needs more citizen scientists to marvel at the workings of forest ecosystems, more communities to save big trees as their most venerable residents, more arbornauts to share their expertise in underexplored places, and more inclusivity of all the diverse voices calling for conservation. If I'm right, with hard work, a whole lot of luck, and a willingness to believe that we're all in this together, one arbornut can seed a forest of citizen scientists ready to stand tall for their beloved trees.

Maybe most important of all, let's teach every kid about the forests that keep them alive. Shall we start by climbing a tree?

GLOSSARY

꘎꘎

accession—in the museum world, to add a specimen to a collection

acute—sharp, or ends with a pointed tip of a leaf

agroforestry—a land-use practice that combines crops with an overstory of native trees

apex—the tip or end

aphid—a small, soft-bodied herbivorous insect that sucks sap from plant stems and leaves

arbornaut—someone who explores the treetops

BioBlitz—a survey of biodiversity, conducted in an organized time frame with a team

biodiversity—the variation of species within a given ecosystem or habitat

biomass—the total weight of living organisms or of one species or individual or community

blade—a flat, green section of any foliage

bract—a scalelike structure that is similar to a leaf but reduced in size

bromeliad—a tropical or subtropical plant with rosettes of stiff, sometimes spiny leaves; some grow as an epiphyte

buttress—a laterally flattened section of a tree trunk base that angles out from the trunk and serves as a type of extension of roots as well as a support mechanism

calyx—a group of sepals making up the outer whorl of floral parts below the corolla

canopy—the upper layer of foliage of a plant or an ecosystem, extending from one inch for moss to three hundred feet for some tropical trees

carabiner—a metal coupling link used in caving or climbing

carnivore—an animal that eats only other animals

carnose—relating to flesh or fleshy

caterpillar—the larval stage of butterflies and moths

cauliflory—flowers and fruits that develop on the tree trunk or older branches instead of on the branch tips as is more typical

cay—a small island of sand or coral rock

chlorophyll—the green pigment in all green plants that is responsible for the absorption of light to supply energy for photosynthesis

chloroplast—a cell containing chlorophyll where photosynthesis occurs

citizen scientists—amateur volunteers who participate in organized research efforts

compound leaf—a leaf divided into leaflike segments called leaflets, with one main stem containing a bud at the base of all the leaflets

cordate—heart-shaped

corolla—a group of petals, whorled and usually the colorful part of a flower, located above the calyx

deciduous—annually shedding leaves, as with many trees or shrubs

decomposer—an organism that breaks down dead matter so it can be recycled

dentate—toothed along the margin, in the case of a leaf

ecosystem—a community of interconnected species plus its physical environment

ecotourism—a low-impact form of tourism that involves visiting pristine or protected areas and follows sustainable practices

emergent—a tree that grows taller than others, so its canopy protrudes above the other trees

endemic—a species that occurs naturally in a specific region

entomology—the study of insects and their relatives, such as spiders, ticks, and mites

epicormic shoots—small branches that grow from the base of trees or trunks, sometimes as a response to stress by the tree whereby the branches are not leafing in a healthy fashion

epiphylly—a layer of microflora growing on the surface of a leaf, usually mosses or lichens

epiphyte—plants that grow on other plants but do not parasitize them

evergreen—a plant that retains green leaves throughout the year

extrafloral nectary—a gland that secretes nectar but is found outside of a flower, such as on a leaf blade, and serves to attract ants to, in turn, protect the foliage from herbivores

extremophile—a microorganism that lives in conditions of extreme temperature, acidity, pressure, alkalinity, or chemical concentration

ficulneus—like or similar to a group of plants of the genus *Ficus* (fig)

frass—pellets of solid waste material from insects

frugivorous—fruit eating

gall—an abnormal growth on plants in response to the presence of insect larvae, or sometimes mites or fungi

genus (*plural* genera)—the first section of a scientific name, which represents a category of several species

glabrous—a leaf surface that is smooth, free from hairs or down

glaucous—a dull gray-green or bluish color

hemi-epiphyte—a plant that spends half its life as an epiphyte and half rooted into the ground

herbarium—a collection of pressed, dried plants with names, locations, and dates of collection

herbivory—the consumption of plant material (an **herbivore** is one who consumes only plant material)

hexaploidy—a cell or nucleus that contains six homologous sets of chromosomes

host specific—a specialized insect feeding on one species of plant

hypothesis—a proposed explanation based on limited evidence as a starting point for further investigation

instar—a phase between periods of molting in the development of an insect

invasive species—a nonnative species that outcompetes native species and/or disrupts a natural ecosystem

jumar—a mechanical clamp attached to a fixed rope that tightens when weight is applied and relaxes when weight is removed

lanceolate—spear-shaped, widest near the base and narrowing toward the tip

larva (*plural* larvae)—an immature stage between egg and pupa; also called grub, caterpillar, maggot

leaf miner—an insect that lives and feeds on the cells between the upper and lower surfaces of a leaf

leaves—the foliage section of any plant

legume—a plant member of the bean family (Fabaceae) that has a fruit similar to a bean pod

LIDAR—Light Detection and Ranging, a laser imaging system used to create high-resolution digital elevation models

lignotuber—an underground root, usually single, that is a major water and nutrient storage unit

mast seeding—highly variable production of fruits by a tree, but a massive fruiting activity that is sometimes annual or as intermittent as decadal

monodominant—when one species of tree predominates an entire forest stand

monoecious—having both male and female reproductive organs in the same individual

mycorrhiza—a symbiotic relationship between a vascular plant's roots and underground fungi, bestowing a competitive advantage by the plant's taking in additional water and soil nutrition

oblong—longer than wide, like an elongated leaf shape

oology—the study of eggs

ornithology—the study of birds

ovary—the part of a flower that develops into a fruit

palmate—a leaf that is shaped like a hand, with usually five to seven multiple lobes like fingers

petal—the colorful part of a flower, often circling around the ovary and pistils

petiole—the narrow, round stem portion of a leaf

phloem—the vascular tissue in plants that conducts sugars and other metabolic products down from the leaves to the roots

photosynthesis—when plants, using the energy from sunlight, make sugar (CH_2O) and oxygen (O_2) from carbon dioxide (CO_2) and water (H_2O)

pistillate—flowers that have pistils but no functional stamens

pistils—reproductive parts of the plant, usually protruding so that pollinators can bring pollen for sexual reproduction of a plant

pooter—a device used to collect small insects by suction, including fruit flies, springtails, ants, and leafhoppers

primary forest—original, pristine forest that is relatively unaffected by human activities

secondary growth—vegetation that replaces the primary or original vegetation

sepal—the lower lobe of the calyx that is usually green and leaflike

shoot—new growth arising from the base of older growth

simple leaf—a leaf not divided into leaflets

species—the second part of a scientific name, representing a single population sharing genetic and structural features

stamen—the male part of a flower made up of a filament or stalk, plus an anther that produces pollen

staminate—flowers with stamens but lacking a pistil

stem—the axis of a vegetative shoot

stigma—the sticky part of the pistil, usually at the top of a columnar receptacle where the pollen sticks and enters the plant for germination

stomata—pores found in leaf surfaces that control the exchange of gases

strangler—a form of fig that begins growing on a canopy branch as an epiphyte but eventually roots into the ground and ultimately forms a woody stem around the host tree where it germinated, usually strangling it

succession—replacement of one type of community by another

sun fleck—a small pocket of sunlight that penetrates through the upper canopy and strikes the lower leaves or ground

tannin—a bitter substance produced in bark and other plant tissues that repels insect attacks

tardigrade—a microscopic organism of the phylum Tardigrada that lives in moist substrates and is related to arthropods; also known as a water bear

taxon (*plural* taxa)—a group of related species that could be family or genus or species

taxonomy—the science of classification and naming of organisms

temperate forest—a stand of trees found in moderate, often four-season climates, located between the tropical and boreal regions

terrestrial—growing on the ground

tropical rain forest—a moist broadleaf forest located close to the equator that has high rainfall and usually, but not always, high diversity

understory—the lower layer of vegetation beneath the main canopy of the forest

vascular bundle—the cluster of vessels or transport tissue located just under the outer bark and containing xylem and phloem tissue

vascular plant—a plant possessing vascular tissue to transport water and nutrients, such as trees, ferns, angiosperms, and gymnosperms, in contrast to mosses

whale's tail—a mechanical device that allows a climber to descend on a rope

xylem—vascular tissue in plants that conducts water and dissolved nutrients upward from the roots toward the leaves

ACKNOWLEDGMENTS

❦

As with most of my international conservation efforts, this book was germinated with the support of a global community, certainly not by me alone. So very many people, as well as thousands of big trees, deserve heartfelt thanks! After traveling and working in forty-six countries over a lifetime, I am grateful for myriad local and international people whose wisdom and friendship were an essential part of my journey. I can never thank everyone on one page, but please know that all of you are illuminated in my heart, soul, and mind with enormous gratitude. A few groups rise to the top: A lifetime of fellow classmates and teachers from Hoffman Elementary School and Elmira Free Academy in Elmira, New York; Williams College in Massachusetts; the University of Aberdeen in Scotland; and in Australia, Sydney University and the University of New England. I especially thank three lifelong friends from kindergarten: Betsy Hilfiger, Mimi Welliver, and Maxine Parker. A cohort of women in outback Australia fostered our amazing sisterhood when we birthed our babies together. Back in the United States, I had some great role models who advised and encouraged me while I struggled to juggle family and work: Tom Lovejoy, Peter Raven, Sylvia Earle, Betsy Bennett, Kristin and John Replogle, Ed Wilson, John and Lee Trott, Hal Heatwole, Carol and Bob Bilbro, Dan Solomon, Jeff Braden, Hugh Caffey, Brian Rosborough, and Cy Spurlino.

In Ethiopia, I thank my soulmate of church forest conservation, Alemayehu Wassie Eshete, plus the priests around Bahir Dar, who always welcomed me with open arms and inspired our collaboration. In Malaysia, the Cockrell family, Alan Tan, and the faculty at the Universiti Sains Malaysia were the local rock stars who ensured our collective success to conserve rain forests in Penang. Soubadra Devy, T Ganesh, and Sinu Pallaty are lifelong friends and catalysts for canopy programs in India. Randy "Waterbear" Miller is one heck of a tardigrade taxonomist, and together we shared our expertise with mobility-limited students. My guides for over twenty-five years in the Peruvian Amazon taught me so much about local uses of plants as well as insights about human nature: Guillermo, Ricardo, Willie, and Juan, to name but a few. My newest initiative, Mission Green, is blessed with a fabulous Science Advisory Board who believe that forest canopies are an integral part of global conservation: E. O. Wilson, Sylvia Earle, Patricia Wright, Wade Davis, Peter and Pat Raven, and Alemayehu Wassie. I was fortunate to work with outstanding staff from different institutions: National Geographic, New College of Florida, Williams College, Marie Selby Botanical Gardens, North Carolina Museum of Natural Sciences, California Academy of Sciences, National University of Singapore, Rolex SA, and the Explorers Club. Dulce Martinez was my Sarasota sister, and Jean Thompson Black was a brilliant mentor who edited several earlier books at Yale University Press.

The Rachel Carson Center for Environment and Society in Munich provided an inspirational writing fellowship, and the staff and board of the TREE Foundation nurtured my storytelling for over two decades. A special "kitchen cabinet" of Amazon travelers—Deb Heineman, Bob Woods, Mary Mountcastle, Cyndie Spencer, Harland Chun, Susan Walker, and Jim Follett—helped germinate Mission Green, as did my Earth sister, Elizabeth Moore, the first angel funder to launch this initiative. My literary agent, Jessica Papin from Dystel, Goderich & Bourret LLC, is herself a force of nature who truly birthed this book with her enthusiasm and influence. The team at Farrar, Straus and Giroux was brilliant and supportive, led by my beloved editor, Jenna Johnson, whose gentle critique turned many rough sentences into a polished draft, and by the ever-ready Lydia Zoells. I am thankful to the FSG rights team: Devon Mazzone, Flora Esterly, Pauline Post, and Anaka Allen; to the production, design, and legal teams: Debra Helfand, Nancy Elgin, Nina Frieman, Abby Kagan, Barbara Cohen, Susan VanHecke, and Na Kim (who made the beautiful illustrations); and to the publicity, marketing, and sales teams: Steve Weil, Sarita Varma, Nikki Barnhart, Hillary Tisman, Daniel Del Valle, Carina Imbornone, and Spenser Lee.

Finally, my sons, Eddie and James Burgess, were an inspirational component of every adventure around the planet, as were my parents, Alice and the late John Lowman, who generously minded the boys during some of my most remote expeditions. And soon, I aspire to take the next generation, grandchildren Lyla Louise and Everett Ely Burgess, to explore the treetops of the world. My extended family—including fellow camp mates from Burgundy Center for Wildlife Studies and fellow graduate students and faculty—all taught me so much during this lifelong leaf quest, ranging from how to swat blowflies, to how to safely capture an enormous python, to how to efficiently analyze reams of data. Big hugs-'n'-bugs to Vic von Klemperer, who cooked amazing cuisine while we read the final proofs, plus other Sarasota friends who served as sounding boards throughout multiple drafts. You know who you are!

Finally, many arbornauts around the world contributed knowledge, inspiration, and perspiration to my stories of trees; I can never thank all of you enough for your dedication to the planet's leafy giants.

INDEX

➤➤⤙⤙